A Guide to Field Guides

Reference Sources in Science and Technology Series

A Guide to the Zoological Literature: The Animal Kingdom. By George H. Bell and Diane B. Rhodes.

Guide to Information Sources in the Botanical Sciences. Second Edition. By Elisabeth B. Davis and Diane Schmidt.

A Guide to Field Guides: Identifying the Natural History of North America. By Diane Schmidt.

A Guide to Field Guides
Identifying the Natural History
of North America

Diane Schmidt

Biology Librarian
University of Illinois at
Urbana-Champaign

1999
LIBRARIES UNLIMITED, INC.
Englewood, Colorado

LIBRARIES UNLIMITED, INC.
P.O. Box 6633
Englewood, CO 80155-6633
(800) 237-6124
www.lu.com

Library of Congress Cataloging-in-Publication Data

Schmidt, Diane, 1956-
   A guide to field guides : identifying the natural history of North America / Diane Schmidt.
    xxvi, 304 p. 17x25 cm.
    Includes bibliographical references (p. 251).
    ISBN 1-56308-707-3 (cloth)
    1. Natural history--North America Bibliography.   2. Animals--North America Identification Bibliography.   3. Plants--North America Identification Bibliography.   4. Rocks--North America Identification Bibliography.   5. Fossils--North America Identification Bibliography.   I. Title.
Z7408.N6S36   1999
[QH102]
016.50873--dc21
            99-22868
            CIP

Contents

Acknowledgments

I wish to acknowledge the Campus Research Board at the University of Illinois at Urbana-Champaign and the Research and Publications Committee of the University Library, which provided travel support for the completion of the research reported in this book. In addition, the staff of the Illinois Research and Reference Center deserves high praise for cheerfully and promptly filling the innumerable Interlibrary Loan requests that I made. Creating this guide required visits to numerous libraries and bookstores across the country from Florida to Alaska, many of which were visited during vacations. And so, to all the friends and relatives who uncomplainingly hung out in libraries while on vacations with me for the past seven years, this book's for you.

Introduction

Why a guide to field guides? For one thing, field guides are extremely popular resources. Almost every American home and library has at least one field guide from the most popular field guide series, and the same is also true for other regions of the world. Besides their obvious utility in helping to identify animals and plants, they are also good sources for illustrations and natural history information as well. In addition, field guides are popular tools in the biologist's tool box, and for many of the same reasons as for amateur naturalists.

What is a field guide? Typically, a field guide is "a book designed to be carried outdoors and used to identify species of birds or other particulars of natural history. A field guide is typically portable, if not 'pocket-sized'; illustrated with diagnostic drawings or photos of the forms under consideration; and filled with concise descriptions of physical details, habits, distribution, and other facts pertinent to identification" (Leahy 1982, 273). A field guide should also be designed for the use of amateurs. There are a number of different technical manuals, atlases, floras and faunas, handbooks, and keys for the use of professionals.

One feature found in many, though not all, field guides is a key. A key is defined as "an arrangement of the salient characters of a group of plants or animals or of taxa designed to facilitate identification" (*Merriam-Webster Collegiate Dictionary* 1993). Most formal taxonomic keys are dichotomous keys, in which the user must select one or the other of two opposing choices (leaves opposite or leaves alternate, for instance), which then leads to another set of choices until the plant or animal is finally identified. There are many other varieties of keys, including tables and visual keys with illustrations or silhouettes of the species to be identified. Although keys generally require some practice to be able to use well, they can be very helpful. Trees in winter, for instance, are more easily identified using a key than the usual field guide descriptions.

While the major field guide series such as the Peterson, Audubon, and Golden field guides are well-known, literally thousands of other field guides have been published, including guides to everything from the scorpions of the American Southwest to the lemurs of Madagascar. There are guides appropriate for every level of expertise, with large print or small print, photographs or illustrations, long descriptions or no descriptions.

Despite the fact that innumerable field guides have been published, it is not easy to discover whether a field guide for a particular group of organisms or region is available. There is no Library of Congress subject heading for field guides. The subheading "Identification" is used for field guides, but it is also used for a number of technical works that are not field guides. Titles are deceptive as well. A book titled "field guide" may, in fact, be a travel guidebook or geological log book, and many field guides do not include those words in their titles.

The field guides annotated in the following chapters are the "normal" field guides commonly found in libraries and bookstores. Due to space limitations, only North American (and Hawaiian) field guides are included. The majority of the field guides covered are in print or are classic guides. A large number of field guides that could not be included in the present volume, particularly international guides and guides in non-standard formats, can be found at the author's field guide web site, at http://www.library.uiuc.edu/bix/fieldguides/main.htm.

HISTORY OF FIELD GUIDES

Field Guide Precursors

Identification guides intended for the use of amateurs have been around for generations. Herbals, for instance, can be seen as ancestral field guides though medieval bestiaries were not. Bestiaries were written as allegorical works, using stories about animals to illuminate Christian morality. Reality has very little place in these bestiaries, and one would certainly not use the stylized illustrations to actually identify the animals represented. Herbals were more reality-based than bestiaries, because their main purpose was in describing the use of medicinal plants. In at least some of the herbals it was clear that the author and artist had actually seen the plants that they were describing and illustrating.

In post-medieval times, the study of nature began to become more "scientific," with less reliance on authority and more on observation and experimentation. The eighteenth century was the heyday of vast, encyclopedic works on plants and animals, such as Georges Buffon's *Histoire Naturelle, Générale et Particulière* (1749–1804), a 44-volume encyclopedia. Its vast size makes it plain that this encyclopedia was not for field use. However, during this same heady period one familiar feature of field guides was established. The first dichotomous key was developed by Jean Baptiste Lamarck, of "inheritance of acquired characteristics" infamy, in the third (1815) edition of his *Flore Française*. The key was designed to help amateurs identify plants, and is instantly recognizable to modern eyes. Like modern keys, it allowed users to identify a plant by following a series of decisions based on the form of the plant.

Natural Theology and the Victorian Age

A major impetus in the development of field guides came with the popularity of nature study in Britain and Europe during the eighteenth and nineteenth centuries. The doctrine of natural theology was first popularized in England by John Ray's *Wisdom of God Manifested in the Works of the Creation* (1691) and later by William Paley's *Natural Theology* (1802). Natural theology made nature study a form of "do-it-yourself theology" (Moore 1993), in which students of the natural world could come to see the mind of God. During the Victorian era especially, nature study was immensely popular. Popular books like Charles Kingsley's *Glaucus; or the Wonders of the Shore* (1873) extolled the wonders of nature and the mind of God displayed therein, but they also assumed that their readers knew a great deal about the natural world.

This knowledge could be obtained from a number of identification guides which were the direct precursors of modern field guides. Kingsley recommended several identification guides in the final pages of *Glaucus*, including the Rev. D. Landsborough's *A Popular History of British Zoophytes, or Corallines* (1852). This little book, not much larger than a modern field guide, described and illustrated British zoophytes (marine invertebrates thought to resemble plants, such as corals, hydras, and sea anemones). While it is fully in the tradition of natural theology, the descriptions seldom wander far from the physical nature of the zoophytes.

Kingsley is not the only author better known for literary works who was involved with natural history. For instance, Beatrix Potter was passionately interested in nature study before she became famous for Peter Rabbit (Lane 1968). She became an expert on fungi in her early twenties, and made numerous paintings of fungi, which she hoped to publish as a scientific work. Unfortunately, they were not considered sufficiently scientific and the book was never published, much to Potter's disappointment. The story has a happy ending, however. More than seventy years after their creation and two decades after Potter's death in 1943, W. P. K. Findlay rediscovered her fungi paintings and created the *Wayside and Woodland Fungi* (1967) field guide around her illustrations.

The American Experience

While natural theology was not unknown in the United States, and nature played an important role in the Transcendentalist movement, nature writing in America focused more on the frontier experience and featured essays rather than moral tracts. The numerous identification guides were full of the excitement of discovery of new species and were very popular. For instance, Thoreau made use of Bigelow's *Florula Bostoniensis* (1824) in his tramps about Concord. Asa Gray's *Manual of Botany of the United States*, first published in 1848, is still one of the standard botanical manuals for the U.S. It was first intended as a pocket-sized manual, but even by excluding everything but keys and descriptions it soon outgrew even the largest pocket.

The American natural history experience is similar to the European experience in its literary connections. For instance, the only one of Edgar Allan Poe's books to be a hit during his lifetime was the semi-plagiarized *The Conchologist's First Book* (1840) (Gould 1995). Famous artists became involved in preparing identification guides as well. While John James Audubon never illustrated an identification guide, Louis Agassiz Fuertes (1874–1927), who is considered by many to be an even better ornithological artist, illustrated several bird books. His illustrations for Ralph Hoffmann's *A Guide to the Birds of New England and Eastern New York ...* (see entry 1064) were a major factor in the book's success and profoundly affected a certain young artist who went on to revolutionize the field.

Early Twentieth Century

By the beginning of the twentieth century, there were a number of identification guides which, like our modern field guides, were designed to be taken into the field. In the United States, for instance, the long-running Putnam's Nature Field Books series (published 1910–1970, see series entry 85) was very popular and covered almost all of the usual field guide subjects. The language in all of these early guides tended to be rather technical, however, and by modern standards the illustrations were often less than useful. The text-to-illustration ratio was also much higher than in modern guides.

Bird guides, for instance, typically contained very detailed descriptions, and identification was supposed to be made with a bird in the hand. They were for the fellow who collected birds with a gun, not a camera or binoculars. Some of the major North American bird identification guides of the times included Frank Michler Chapman's *Handbook of Birds of Eastern North America* (see entry 1052) and F. Schuyler Mathews' *Field Book of Wild Birds and Their Music* (1904; see entry 1071, which compared bird songs to, among other things, themes from opera!). The closest thing to our modern field guides were the two-volume pocket sized *Bird Guides* written by Chester A. Reed (1909; see entry 1087), which were hugely popular and sold for $1.00 in the thirties.

Roger Tory Peterson

Anyone who knows anything about field guides knows what comes next. In the early 1930s a young artist by the name of Roger Tory Peterson had a bright idea for a simplified guide to the field identification of eastern birds. He managed to find a publisher willing to gamble on the chance that Depression-era bird lovers would pay for a new and different kind of bird book, especially one that cost a whopping $2.75. The gamble paid off. The first edition sold out its 2,000 copy first printing within three weeks, and has been a steady seller ever since. In 1986, the first edition was reported to be worth $300 and up. One copy apparently was sold for $1,200 (Cassie 1986).

Peterson's impact on the hobbies of birding and nature study is well known (just look at any of his obituaries), but what is sometimes difficult to comprehend is how thoroughly revolutionary his 1934 bird book was. Early identification guides were far less uniform than modern guides, and, while most aspects of *A Field Guide to the Birds* can be found scattered throughout other guides, Peterson was the first to put all of them together. For instance, prior to *A Field Guide to the Birds*, identification guides were generally not called field guides. The only recently found book titled "field guide" published before 1934 is Nininger's *Field Guide to the Birds of Central Kansas* (1927), which in modern terms is a checklist rather than a field guide. However, as early as 1925 some of the Cleveland Museum of Natural History's Pocket Natural History series (see series entry 81) were described as field guides by the publisher and do fit the modern definition.

Peterson is best known for his illustrations. Prior to *A Field Guide to the Birds*, illustrations often showed only one or two species per page, generally in a natural setting. Peterson's innovation was to paint simplified illustrations of the birds, showing only those aspects that would most reliably enable the user to identify a bird at a distance (the "field marks") . He got that idea from a chapter in Ernest Thompson Seton's book *Two Little Savages* (1903), in which a boy trying to identify ducks gets the idea of sketching them as they would appear from a distance. Seton, an accomplished artist, included appropriate illustrations that are strikingly similar to Peterson's duck illustrations in the first edition.

Peterson's illustrations became less schematic in later editions, though the field marks are still indicated. Birds were also grouped by appearance, with several easily confused species grouped together in a single plate. It is interesting to note that some recent field guides have gone back to the older method of showing just a few species per plate in naturalistic settings, and then calling this a modern innovation. The "Peterson System" also includes arrows pointing to the most important field marks in the illustrations. While most modern field guides use simplified illustrations and brief descriptions of field marks, not all include the arrows.

Often overlooked in discussions of Peterson's innovations is his language. Instead of describing a bird in exhaustive detail, Peterson pointed out only the features of appearance or behavior that allowed an observer to distinguish between species. His terminology was descriptive rather than technical. Peterson himself complained that earlier guides, such as Chapman's (1917) *Handbook of the Birds of Eastern North America*, went into great descriptive detail while overlooking obvious identification points (Peterson 1983). Chapman's description of the robin, for instance, provided a lengthy description of the bird's coloration, almost literally from head to foot but never pointed out that the robin is the only North American bird with an orange belly. His descriptions of other birds do highlight salient points more clearly, however. Conciseness can be taken too far as well. In the first edition of Peterson's *A Field Guide to the Birds*, the robin is not illustrated at all, and its description reads, in full, "The one bird that everybody knows" (Peterson 1934, 107). By the fourth edition, this was amplified to read, in part, "A very familiar bird . . . Recognized by its dark gray back and brick-red breast" (Peterson 1980, 220).

The value of Peterson's prose was recognized at an early stage. Elizabeth Bishop, writing to her friend and fellow poet Marianne Moore, said that she had just received a copy of *A Field Guide to the Birds* (probably the second edition), and found the illustrations valuable. However, the poet also remarked that "the descriptive writing is quite good and different, too (Bishop 1941)." In 1996 Copple enthused about Peterson's painterly and poetic language. The poetry lies primarily in Peterson's choice of color terms and in his descriptions of songs and calls.

Success of Field Guides

Following the success of Peterson's first field guide, Houghton Mifflin asked Peterson to create a series of guides based on his field mark system. Peterson was the author or artist for several guides and edited the remaining volumes. The series rocketed to success and spawned numerous competitors and imitators.

The popularity of field guides as a genre can hardly be overstated. While sales figures are often untrustworthy, they can be quite interesting (see Table 1). In 1987, the annual sales figures for all field guides was 750,000 per year; in 1996, Peterson's *Field Guide to Eastern Birds* alone sold about 100,000 copies annually. Many field guides remain in print for years and are reliable backlist titles for publishers large and small.

Another measure of the popularity of field guides is their secure place in popular culture. Field guide authors such as Roger Tory Peterson are famous, and one even provided the name for Ian Fleming's best-known creation (Bond 1966). James Bond was the author of the classic *Birds of the West Indies* (1936). This field guide was well known to Fleming, who was living in Jamaica when he wrote the first 007 thriller. Fleming was looking for a short, nondescript name for his then un-sexy hero and liked the sound of "James Bond." Few of us would pick James Bond as a nondescript name these days.

Field guides have also become popular enough to attract parodies. The Sills, for instance, have lovingly spoofed bird field guides in their *Another Field Guide to Little-Known and Seldom-Seen Birds of North America* (see entry 1030) and its sequel, inventing plausible new species and poking fun at some of birders' less-noble traits. The field guide format has also been used to poke fun at the North American Female's pursuit of the North American Male (see entry 1179). There are field guides to cat poses (see entry 1185), roadkill (see entry 1182), windshield debris (see entry 862), and alas, much more.

"Field guide" is a generic term these days, used for a thousand purposes, most of them having little to do with the identification of plants or animals. There are articles claiming to be field guides to the United States economy, southern men, bishop watching, cyberspace, and much, much more. Authors use the term as a catchy attention-grabber with little reference to any of the features of a "real" field guide as defined above.

TYPES OF FIELD GUIDES

Print Guides

While the history of bird field guides is the best known and most frequently reported, identification guides to most other plant and animal groups have long been available though they may show different influences than the bird guides. For instance, numerous tree guides were published by federal and state agriculture departments or experiment stations due to the economic importance of timber. Mushroom guides emphasize the edibility of species; seashell guides the collectability of shells.

Since field guides are so popular, many types of publishers have gotten into the business. Not only are large commercial publishers such as Houghton Mifflin in the business, but also university presses such as Princeton University Press and small presses such as Lone Pine. A number of specialized field guides are published by state and national government departments and may be found at depository libraries. Local field guides are frequently published by small presses and sold locally to libraries and bookstores. An incredible range of types of field guides have been created by this range of publishers; almost every conceivable type of organism or object is covered for almost every region of the world, and almost every level of expertise is catered to.

The authors of field guides range from well-known experts with full academic credentials to fanatical amateurs, some quite proud of their lack of academic degrees. The author's lack of credentials does not automatically lessen the value of a field guide, since many amateurs know their local flora or fauna as well as any "perfesser." Roger Tory Peterson, an artist by training, is the most notable example of this group. His rivals in the bird field guide series market, S. Chandler Robbins (Golden Field Guide) and Stephen Farrand (Audubon), are both ornithologists, however.

Many guides are published for the benefit of children and beginners. The familiar National Audubon Society Pocket Guide, Golden Guide, and Peterson First Guide series are all widely available and cover a few major species for beginners. Many guides published by children's publishers and aimed at school and public libraries are also available, though they are generally not annotated in this guide. Beginner's guides aimed at adults are numerous. The numerous beginner's guides to bird-watching, for instance, include information on how to spot and identify birds as well as the usual illustrations and descriptions.

Another field guide category is the souvenir guide for tourists. These tend to be colorful, inexpensive pamphlets covering the gaudiest or most common plants or animals of a popular tourist location (the Caribbean or Hawaii, for instance). The souvenir guides may also include local folklore and similar snippets of natural history interest. They may be of little value to residents of the area with a serious interest in nature, but are often a useful introduction for visitors. For the most part, these tourist guides are not annotated in this book.

On the other end of the spectrum, there are a number of advanced guides. Many of these advanced guides are for serious birders, though they can be found for almost all groups of organisms. They assume a far greater level of expertise than most guides, and are often created for difficult groups or for identifying organisms in far greater detail than other guides. An advanced guide to seabirds, for instance, might allow the user to identify which molt a juvenile is undergoing (and thus its age).

Technical identification guides intended for the use of students or professional biologists are also common. They are generally more difficult to use than field guides but may be useful for amateurs. Manuals and handbooks are guides to the identification of plants or animals, and the terms are often used interchangeably. Originally, a handbook was similar to a field guide because it could fit in the hand, but this distinction is now lost and handbooks often run to many large, heavily-illustrated volumes. Field manuals are fairly technical works designed to be used in the field, and are generally for students.

There are a growing number of taxonomic guides which include birds from around the world but are limited to just one or two closely related families. Most of these are fairly hefty publications and are rather expensive compared to other field guides. Some include extensive information about natural history, but all cover identification in great detail and have lengthy bibliographies. These taxonomic guides are useful for specialists and libraries, but are of less use as field guides for the average birder. While they may be excellent resources for confirming an identification or obtaining information about habits or habitats, they are too bulky and expensive, and cover too many locations, to be useful for most people. Princeton University Press, Yale University Press, and Houghton Mifflin in the U.S. and Pica Press and Christopher Helm in the U.K. have all published or co-published numerous taxonomic guides that are almost identical in format. These taxonomic guides are not included in this annotated guide because they are not true field guides, but an extensive list can be found at a companion Web site, http://www.library.uiuc.edu/bix/fieldguides/main.htm.

The vast majority of field guides assume some degree of experience and are aimed at adults but are easily accessible for motivated children. Most field guides also cover a fairly extensive geographical area, from an entire continent or country to a biogeographical region such as the Western United States. Country-wide or state-wide guides also exist, but political borders seldom make any sense in biological terms so these guides are useful in surrounding areas as well.

Because field guides are designed to be used outdoors, often in inclement weather, there is a growing sub-set of weatherproof guides. A good, useful guide should be sturdy and withstand a little dampness, but some guides take this much further. The National Audubon Society Field Guide series, for instance, has plastic covers, and a few smaller field guides are printed on Mylar-type plastic pages. Another group of waterproof guides is laminated or plastic cards covering a few popular species. For instance, there are fish-watcher's guides that are printed on plastic and designed for use underwater by snorkelers or scuba divers. Laminated guides for landlubbers are increasingly popular as well. These cards are generally no larger than 8½x11 inches, though the Peterson FlashGuide series and the Pocket Naturalist series from Falcon Press consist of folded, laminated cards that are quite large when unfolded. Since the laminated guides are generally not found in libraries, they are not annotated here but a list can be found at http://www.library.uiuc.edu/bix/fieldguides/main.htm.

Multimedia Guides

A growing number of guides are being created in multimedia or audiovisual formats. There have been videos and sound recordings for identifying birds and bird songs for many years. The growing popularity of personal computers has guaranteed that CD-ROM guides could not be far behind, and, in fact, there are several bird guides on CD-ROM, including the Peterson and Audubon guides for North America.

CD-ROMs and videos, of course, are old hat these days, and the World Wide Web is cutting-edge technology. Naturally, there are some field guide-type resources available on the Web. For instance, the Macphail Woods Ecology Forestry Project on Prince Edward Island in Canada (http://www3.pei.sympatico.ca/garyschneider/guides.html) has posted online field guides to shrubs, birds, trees, wildflowers, and amphibians on their Web site. Another Web field guide is the electronic version of Fautin's *Field Guide to Anemone Fishes and Their Host Sea Anemones* (see entry 872), which is available at the University of Kansas's Biodiversity and Biological Collections Web Server (http://www.keil.ukans.edu/ebooks/intro.html). Such multimedia and/or electronic resources have not been extensively researched for this guide, because the emphasis is on identification guides that can be taken into the field. While these multimedia products are useful resources for home study, they are impractical for field work. The technology will have to improve drastically before electronic media become viable tools for field identification.

ABOUT THIS BOOK

The Annotations

The annotations in this guide are concise descriptions of the field guides, with little evaluative commentary. There are at least two reasons for the lack of evaluation. First, the choice of a field guide is a personal one. The choice of photos over illustrations, or broad or narrow geographical regions, depends on the user's needs and experience. A number of articles evaluating the major field guides have failed to reach a consensus. (For example, see the Annotated Bibliography.) Also, the accuracy and reliability of field guides for regions or groups for which the annotator has no direct experience cannot easily be determined. Many major guides are reviewed by experts in various publications, so the reader wanting authoritative opinions should check for reviews. Reviews can be found both in scholarly journals such as *Auk* and in popular magazines such as *Natural History*.

All field guides annotated here have been personally examined by the author, unless otherwise indicated. The emphasis is on North American field guides that are available in the United States, either guides that are held in libraries or are available in bookstores. While some classic works are included, the emphasis is on field guides that have been published after 1980 and are still in print. Numerous international field guides held in libraries in the United States that were discovered while preparing this book can be found at the author's "International Field Guides" Web site at http://www.library.uiuc.edu/bix/fieldguides/main.htm.

Format of Annotations

The annotations have a standardized format. The number of species covered, the presence of a key, the type of illustrations, range maps, the arrangement of species accounts, and the presence of natural history or introductory materials are listed in a concise format. These brief annotations are followed, where needed, by more general descriptive and bibliographic information. Illustrators are also listed, if two or fewer illustrators were responsible for all illustrations and were credited. Photographers are not listed, even where known. The lack of additional notes in an entry does not indicate that a field guide is at all substandard, just that it follows the general pattern of field guides or is a standard entry in a series.

Series

Because many field guides are published in a standard format in series, these series are described in more detail (and with more evaluation) in Chapter 1. This chapter should be checked for more description of series titles. In some cases, it was not possible to see all volumes in a series. In these cases, the field guide annotation simply refers back to the series description in Chapter 1.

Arrangement of Annotations

The field guides are arranged by type of organism and region covered. General guides that cover a number of organism groups are listed in separate chapters, "Flora and Fauna" for ecosystem-wide guides including both plants and animals, and "Plants" or "Animals" for guides that include groups of organisms from more than one category or which do not belong in another chapter. Chapter subjects were selected according to the number of field guides written about those groups of plants or animals. Thus, marine invertebrates are covered in a separate chapter, but non-insect land invertebrates are included in the chapter titled "Animals."

The chapters are subdivided by region, beginning with field guides for North America as a whole (including Mexico), followed by field guides for eastern and western North America. The dividing line between east and west is the eastern edge of the Rocky Mountains. Thus, the Great Plains states and Texas are considered part of the eastern United States. The Rocky Mountains also form the dividing line for Canadian guides. Field guides for Hawaii are found in the western section.

Caveat

While this book has been in process for eight years and over 75 libraries and bookstores have been visited in most regions of the United States, not to mention searches done in many databases and innumerable Interlibrary Loan requests made, many field guides were inevitably overlooked. In some cases, books that claim to be field guides were excluded because they did not fit the definition of a field guide quoted earlier. In many more cases, the field guide was simply overlooked. There is also no end in sight of the proliferation of new field guides, so of course many recent titles missed the publication deadline. If a field guide is not listed here for the topic or area you are interested in, that does not mean it does not exist. Check the field guide Web page listed above. If that doesn't work, try your local library. And, happy hunting!

Additionally, every effort was made to locate International Standard Book Numbers (ISBNs) for each title in this book, but in some cases it was not possible to find one.

Table 1.
Sales of Field Guides

Title/Series	Sales figures	Year	Source
All field guides	750,000/yr 600,000/yr	1987 1986	Culhane Lipske
All Peterson guides	18 million 9 million	1996 1983	Leo Appelbaum
Audubon series (bird guides only)	Over 1 million	1983	Appelbaum
Golden Field guide, Birds of North America	5 million	1996	Lipske
Golden guides, Birds	7 million	1983	Appelbaum
Peterson, Birds of Britain and Europe	750,000 (13 languages)	1977	Devlin and Naismith
Peterson, Animal Tracks	400,000	1990	Graham
Peterson, Eastern and Western Birds	7 million combined	1996	Adler
Peterson, Eastern Birds	5 million (100,000/yr)	1996	Gordon
Peterson, Wildflowers	1.25 million	1996	Adler
Simon and Schuster guides	850,000	1983	Appelbaum

REFERENCES

Adler, Jerry. 1996. The original field guide. *Newsweek* 128 (August 12): 60.

Appelbaum, Judith. 1983. Pursuing the wild whatever. *New York Times Book Review* 39–40 (September 25, 1983).

Bigelow, Jacob. 1824. *Florula bostoniensis.* 2nd ed. greatly enlarged. Boston: Cummings, Hilliard, and Co.

Bishop, Elizabeth, to Marianne Moore. December 28, 1941. *Elizabeth Bishop: One art,* ed. Robert Giroux. (New York: Farrar, Straus, Giroux, 1994), 104.

Bond, James. 1936. *Birds of the West Indies.* Philadelphia: Academy of Natural Sciences of Philadelphia.

Bond, Mary Wickham. 1966. *How 007 got his name.* London: Collins.

Buffon, Georges Louis Leclerc, compte de. 1749–1804. *Histoire naturelle, générale et particulière, avec la description du Cabinet de roi.* Paris: De l'Imprimerie Royale. 44 v.

Cassie, Brian E. 1986. Collecting the first "Peterson" bird guides. *AB Bookman's Weekly* 78: 355–356.

Chapman, Frank M. 1917. *Handbook of the birds of eastern North America.* Rev. ed. New York: D. Appleton and Co.

Copple, Nathan. 1996. A field guide to the words . . . of Roger Tory Peterson. *Living Bird* 15(1): 30–33.

Culhane, John. 1987. The hobby that lifts your heart. *Reader's Digest* 130 (February): 56–60.

Devlin, John C., and Grace Naismith. 1977. *The world of Roger Tory Peterson: An authorized biography.* New York: Times Books. pp. 106–107.

Findlay, Walter Philip Kennedy. 1967. *Wayside and woodland fungi.* London: F. Warne.

Gordon, John Steele. 1996. Inventing the bird business. *American Heritage* 47(8): 18.

Gould, Stephen Jay. 1993. Poe's greatest hit. *Natural History* 102(7): 10–19.

Graham, Frank Jr. 1990. Mardy at 88. *Audubon* 92(6): 12–14.

Gray, Asa. 1848. *A manual of the botany of the northern United States from New England to Wisconsin and south to Ohio and Pennsylvania inclusive arranged according to the natural system*. Boston: Munroe.

Kingsley, Charles. 1873. *Glaucus; or, The wonders of the shore*. 5th ed. London: Macmillan.

Lamarck, Jean Baptiste Pierre Antoine de Monet de. 1815. *Flore française, ou descriptions succinctes de toutes les plantes qui croissent naturellement en France . . .* 3rd ed. Paris: Desray.

Landsborough, D. 1852. *A popular history of British zoophytes, or corallines*. London: Reeve.

Lane, Margaret. 1968. *The Tale of Beatrix Potter*. Rev. ed. London: Frederick Warne.

Leahy, Christopher. 1982. *The birdwatcher's companion: An encyclopedic handbook of North American birdlife*. New York: Hill and Wang.

Leo, John. 1996. He was a natural. *U.S. News and World Report* 121(6): 17.

Lipske, Mike. 1986. They're not called birdwatchers anymore. *National Wildlife* 24(October/November): 46–51.

Mathews, Ferdinand Schuyler. 1904. *Field book of wild birds and their music: A description of the character and music of birds, intended to assist in the identification of species common in the United States east of the Rocky Mountains*. New York: G. P. Putnam's Sons.

Moore, John A. 1993. *Science as a way of knowing: The foundations of modern biology*. Cambridge, MA: Harvard University Press. p. 138.

Nininger, Harvey Harlow. 1927. *A field guide to the birds of central Kansas*. McPherson, KS: Democrat-Opinion Printer.

Paley, William. 1802. *Natural theology, or, Evidences of the existence and attributes of the Deity: collected from the appearances of nature*. London: Printed for R. Faulder.

Peterson, Roger Tory. 1934. *A field guide to the birds: Giving field marks of all species found in eastern North America*. Boston: Houghton Mifflin.

Peterson, Roger Tory. 1980. *A field guide to the birds: A completely new guide to all the birds of eastern and central North America*. 4th ed. completely rev. and enl. Boston: Houghton Mifflin.

Peterson, Roger Tory. 1983. Books of a feather. *National Wildlife* 22 (December 1983/January 1984): 22–28.

Poe, Edgar Allan. 1840. *The conchologist's first book: A system of testaceous malacology* . . . 2nd ed. Philadelphia: Published for the author by Haswell, Barrington, and Haswell.

Ray, John. 1691. *The wisdom of God manifested in the works of the creation.* London: Printed for Samuel Smith, at the Princes Arms in St. Paul's Church-yard.

Reed, Chester Albert. 1909. *Bird guide.* New York: Doubleday, Page and Co.

Seton, Ernest Thompson. 1903. *Two little savages; being the adventures of two boys who lived as Indians and what they learned.* New York: Doubleday, Page and Co.

Field Guide Series

Many field guides are published in series. Some of the series have strong editorial control and are very uniform, while the volumes in other series have little in common beyond the series name. Not all series are annotated in this chapter, but an attempt has been made to include all major series and all series that include volumes listed in the field guide annotations, but which were not seen. The series descriptions include an indication of the range of subjects covered, the type of illustrations, the level of expertise required, and occasionally information on the publisher. The series in this chapter are arranged alphabetically by series name. There are some single-author series that have very similar titles and formats but do not have an official series statement. These are alphabetized by the author's name and/or common title words (for example, series entry 3).

1. **Alaska Pocket Guide.** This is a series of small field guides published in the latter half of the 1990s. They cover the most common plants and animals of Alaska. The attractive volumes feature color photos. Subjects include mammals, mushrooms, fish, and birds.

2. **American Nature Guides.** This series of bargain-priced spiral-bound guides feature color illustrations and a tabular layout. Each species is covered on a single page. Dragon's World publishes the same or similar titles in Britain.

3. **Animal Tracks** series. Ian Sheldon has written several guides to animal tracks, each volume covering 40–60 mammal species plus 5–10 birds and reptiles. They include illustrations of both tracks and animals, plus natural history and a discussion of similar species. There is also an "Identification Chart" comparing and identifying tracks. This series is slightly larger than Stall's *Animal Tracks* series (see entry 4), but is still pocket sized.

4.   **Animal Tracks** series. Chris Stall has written several pocket-sized regional guides illustrating the tracks of North American animals. Each pocket sized book covers 30–40 species of mammals plus a few birds, reptiles, and amphibians. The descriptions are fairly extensive and include natural history, but there are no illustrations of the animals themselves, only their tracks.

5.   **Arizona Traveler Guidebooks: American Traveler Series**. A series of several small pamphlets covering the geography and natural history of Arizona. Similar to the *Colorado Traveler Guidebooks*, below.

      ❑ **Audubon Field Guides**. See National Audubon Society, entry 63.

6.   **Audubon Society Beginner's Guide**. Not to be confused with the *Audubon Society Pocket Guides* (see entry 8), this is a series of very small children's guides, each covering about 50 common plants or animals. Unlike the other Audubon Society guides, they feature color illustrations rather than photos. Most were written by George S. Fichter.

7.   **Audubon Society Nature Guides**. A series of guides covering all the major species of plants and animals for a particular ecosystem. Like the other Audubon guides, they feature attractive color photos separated from the species accounts. The species accounts include quite a bit of natural history. They are particularly useful for travelers venturing into a new region, but due to the wide range of groups covered, they can only include the most common species. Regions in the series include deserts, both the Atlantic and Pacific coasts, eastern and western forests, and wetlands.

8.   **Audubon Society Pocket Guide**. This series of pocket-sized beginner's guides, like other Audubon Society guides, features color photos. Most cover 80 species of North American plants and animals, each species on a two-page spread. In addition to the full-page photo, each species account includes natural history, distribution, and may list similar species. The series also includes attractive little volumes on the weather, constellations, and other non-living objects and phenomena.

9.   **Backpacker Field Guide series**. A very recent series of compact field guides published by Explorer Press, at present limited to three volumes by George Petrides on the trees of the western United States. The tree guides are similar in format to Petrides' Peterson Field Guide series books on trees (see entries 474, 474, 511).

10. **Barron's Mini Fact Finders**. This series of pocket-sized guides was originally published in German by Gräfe und Unzer as the GU Kompass series. Some volumes have also been published in French by Nathan as part of the Guide Naturaliste series.

11. **Birds of . . .** This series of city or regional bird guides is written by Chris C. Fisher and published by Lone Pine. Each covers about 125 species and feature color illustrations. The descriptions include similar species, a Quick ID section, and a chart showing the relative abundance of the species throughout the year. Some volumes also include a checklist. They are written for the beginning birder.

12. **California Natural History Guides**. This series currently numbers about 40 guides, most published in the 1970s. Titles include *Mammals of the San Francisco Bay Region* and *Seashore Plants of Northern California*. Since the volumes in this series cover rather narrow subjects and regions, they generally include extensive introductory and background information, as well as natural history. Most volumes have black-and-white illustrations, with a few color illustrations or photos in separate plates. Most also include checklists. They are small enough to carry along on a hike. The authors are generally well-known authorities.

13. **Christopher Helm Identification Guides**. A series of bird handbooks published in the United Kingdom by Christopher Helm and in the United States by Houghton Mifflin and Princeton University Press. Each volume in this highly uniform series is written by well-known experts and has color illustrations by leading bird illustrators. The illustrations are in a separate identification section with brief descriptions and identification tips on facing pages. The identification section is followed by species accounts offering extensive information on the natural history and identification of the birds. Each volume covers a different family of birds worldwide. They are not true field guides, because they are too heavy to carry in the field and are worldwide, but advanced birders find them extremely useful.

14. **Collins Complete Guides**. This group of guides covers all the major groups of plants and animals for a region. Because they must cover a very large number of species, only the common ones are included and the color illustrations tend to be rather small. Like other comprehensive guides, they are useful for people who don't want to burden themselves with several guides but they can't include all species in each group.

15. **Collins Field Guides**. Collins (now HarperCollins) is the major publisher of non-Western Hemisphere field guides, and has several series. Many are also published in French by Delachaux et Niestle. The recent HarperCollins guides are easily recognized by their white covers with a red banner across the top. The Field Guide series covers the major groups of plants and animals in Britain and Europe, with color illustrations and natural history. Some volumes have range maps, and the illustrations may either be in separate plates or integrated with the text. Includes volumes on birds, rare birds, mammals, reptiles and amphibians, insects, butterflies, and trees.

16. **Collins Gem Guides**. This series consists of small, pocket-sized guides to British and European plants and animals. They may have color photos or illustrations. Despite the small size, however, they are not specifically designed as children's guides.

17. **Collins Guides**. Contain more background and introductory information than the Collins Field guide series described above, and are aimed at beginners. Titles include Irish birds and others.

18. **Collins Handguide Series**. This series features a compact format illustrating common species. Text and illustrations are integrated, but unlike most field guides the text describing the individual species is run together. Illustrations generally show the plants or animals in a naturalistic setting. Titles in the series include *Wild Flowers of Britain and Northern Europe and Trees of Britain and Northern Europe*.

19. **Collins How to Identify**. "A new series, Collins How to Identify guides are designed to help even those with the most rudimentary knowledge identify different species with confidence." The series features extensive background information.

20. **Collins Illustrated Checklist**. The volumes in this new series are essentially normal field guides, except that they cover a large number of species and have briefer descriptions and smaller illustrations than most Collins guides.

21. **Collins Nature Guide**. This series features smaller volumes with color photos and brief descriptions. Some were originally published by Gräfe und Unzer in Germany.

22. **Collins Photo Guides**. This series features color photos a la the Audubon series. There are numerous titles in the series, including *Orchids of Britain and Europe*. The text and photos are generally integrated.

23. **Collins Pocket Guide**. Despite the title, volumes in this series are the standard field guide size and feature color illustrations. Subjects include the stars, the seashore, wildflowers, birds, and much more; most cover British or European subjects and species.

24. **Collins Safari Guides**. "This brand new series is designed to give everyone visiting far away places a pocket guide to the animals that they might see." Volumes to date include larger mammals of East Africa and birds of East Africa.

25. **Collins Watch Guides**. These guides are designed to introduce various aspects of natural history to children. Each book has color and black-and-white illustrations, is laminated with a spiral binding, and has fold-out pages for quick identification. The series was originally published in French by Gallimard Jeunesse as Carnets de la Nature.

26. **Collins Wild Guide**. A pocket-sized series of beginner's guides to the plants and animals of Europe. The descriptions are in the form of an "Identification Fact File," that provides heading for each major part of the organism (flower, leaf, stem, etc.). Topics include birds, butterflies, flowers, mushrooms, and trees. All were published in 1996.

27. **Collins Wildlife Trust Guide**. A new HarperCollins series, begun in 1998. The volumes feature color photos and are supported by the U.K. Wildlife Trusts. The bird guide (the only one seen at the time of writing), is a reprint of an earlier Collins guide and is modeled after the National Audubon Field Guide Series (see entry 63).

❏ **Colorado Traveler Guidebooks; American Traveler Series**. See Arizona Traveler Guidebooks, entry 5.

❏ **Concise Guides** (New Holland). See Green Guides, entry 43.

28. **Corrie Herring Hooks Series**: A series of books published by the University of Texas Press. They seem to have little in common apart from the series title and the general theme of natural history, since they include the U.S. publication of the Collins New Generation guides, birder's guides, and much more.

29. **Costa Rica Field Guides**. A series of laminated cards covering the flora and fauna of Costa Rica. At present, five bird and one mammal cards have been published. Other guides that have been announced

include *Wildlife of the Caribbean Slope, Pacific Coast Wildlife, Orchids of Costa Rica, Birds of the Osa Peninsula, Birds of Carara, Game Fishes of Costa Rica, Diver's Guide to Costa Rica, Butterflies of Costa Rica,* and *Venomous and Non-Venomous Snakes of Costa Rica.* Each guide covers 50–100 species, with full color illustrations of the animals in natural poses. The only information provided for each species is the English, Spanish, and Latin names, plus length in centimeters. A portion of the purchase price goes to support conservation programs in Costa Rica.

30. **Discovering Sierra Series**. A series of publications put out by the Yosemite Association, dealing with the plants and animals of Yosemite National Park.

31. **Easy Field Guides**. Most of these small pamphlets were written by Dick and Sharon Nelson. The earlier ones were published by Tecolote Publishers of Glenwood, New Mexico, and later ones are from Primer Press of Phoenix, AZ. The guides cover a number of plant and animal groups of the desert Southwest and the Glacier National Park area. There are black-and-white illustrations.

32. **Easy Way to . . . Recognition**. This trio of books by John Kilbracken features book-length keys written in very simple language for children or beginners. They were published by Kingfisher and have color illustrations and checklists. Topics include trees, flowers, and birds of the U.K. The bird book received a Times Educational Supplement Information Book award.

33. **Ecotravellers' Wildlife Guides**. This recent series, published by Academic Press, is planned to cover the most popular ecotourist destinations in South and Central America and Hawaii. The volumes include descriptions of parks and nature preserves, extensive natural history, and color plates identifying common amphibians, reptiles, birds, and mammals. The emphasis is on the natural history of the region covered.

34. **The Elma Dill Russell Spencer Foundation Series**. A series of otherwise unrelated books sponsored by the foundation. All cover various aspects of Texas natural and human history.

35. **ETI World Biodiversity Database** CD-ROM series. This series of CD-ROM programs features a mix of popular and esoteric subjects. They are published by Springer Verlag for the Expert Center for Taxonomic Identification, a non-profit organization sponsored by the government of the Netherlands, the University of Amsterdam,

and UNESCO. Some of the databases are similar to field guides, particularly the *Birds of Europe* and the *Marine Mammals of the World*. Some are based on other print works while others are new for the series. While they are not very useful in the great outdoors, they are a good source of information and illustrations. All include glossaries.

36. **Eyewitness Handbooks**. This series, published by Dorling Kindersley, covers a variety of topics from shells to the cheeses of France in attractive, slightly oversized volumes. They are not true field guides, since most volumes cover 500 species of plants or animals worldwide. It is not very useful to have a guide to 500 bird's eggs of the world when you are in the U.S. which has at least 600 species of breeding birds. On the other hand, they are very attractive and serve well as a source of color photos and brief "bios" of their subjects.

37. **Falcon Field Guide**. This series includes several books identifying edible berries of regions of North America. The books include recipes. Falcon Press publishes and distributes numerous other field guides that are not identified as part of this series.

38. **FAO Species Identification Field Guides for Fishery Purposes**. The Food and Agriculture Organization of the UN publishes several field guides that identify the marine species of major economic importance to Third World regions. The volumes may have sections on seaweeds, marine invertebrates, fishes, sea turtles, seabirds, and marine mammals. They feature visual keys to the families of plants and animals discussed and have black-and-white illustrations, though some have color photos in separate plates. Written in English, each description includes nomenclature (other scientific names), approved common names in English and other European languages such as Spanish and French, local names, size, habitat, and fisheries notes.

39. **FAO Species Identification Sheets for Fishery Purposes**. This series is similar to the FAO Species Identification Field Guides for Fishery Purposes series described above. Some are in looseleaf format, though the more recent volumes are paperback or spiralbound.

40. **Golden Field Guide Series**. These are the full-scale field guides published by Golden Press. Unlike the Peterson and Audubon series, the Golden Field Guides each cover all of North America. They feature color illustrations integrated with the species accounts and range maps. Titles include birds, trees, rocks and minerals, seashells, and amphibians.

41. **Golden Guides**. This series covers almost every topic imaginable, from fungi to birds to Native American arts to the weather. They are truly pocket sized, and each covers about 200 species with color illustrations, range maps, and natural history. They are designed for children, but are useful for anyone who wants a general, inexpensive introduction to the most common varieties of a group.

42. **Golden Regional Guides**. This now-defunct series was published in the late 1950s and 1960s, and were pocket-sized guides to a region, including flora, fauna, geology, history, and tourist attractions. They are of primarily historical interest at the present, due to their age. Most volumes covered regions of the United States, but volumes on Mexico and Israel were also published. Most of the authors and illustrators were associated with other Golden guides.

43. **Green Guide (New Holland)**. A pocket-sized series featuring color illustrations and brief descriptions of common organisms of Britain and Europe. Originally published, also by New Holland, as Concise guides.

44. **Guide d'Identification Fleurbec**. A series of field guides in French covering the flora of Québec, each with color photos. Most were edited by Groupe Fleurbec, headed by Gisele Lamoureaux.

45. **Guide to Florida Wildlife and Nature**. A series of small pamphlets published by various small presses in Florida, written by Robert Anderson. They feature black-and-white illustrations and cover most topics in Florida's natural history.

46. **Gulf's Field Guide Series** (formerly Texas Monthly Field Guide series). Now published by Gulf Press and formerly by Texas Monthly Press. Most cover Texas or Southwest areas, though the series has recently expanded to field guides of California and Florida.

47. **Illinois Natural History Survey Manual**. This series of pocket-sized volumes is produced by the Illinois Natural History Survey. Most were published from 1936–1992. Topics include Illinois land snails, freshwater mussels, mammals, wild flowers, and shrubs. Most feature either black-and-white photos or illustrations and use fairly technical language.

48. **Instant Guides**. A pocket-sized series of bargain guides. Most include checklists, small illustrations, maps, and an attractive layout.

49. **Instant Nature Guides**. Several children's guides published in 1979, with color photos or color illustrations and a full-color dial on the front and back covers with color illustrations and page numbers of species accounts. All were written by Riki Kondo and illustrated by Dorothea and Sy Barlowe.

50. **Interpreting the Great Outdoors**. This oversized series of books, all but one written by Beverly Magley and illustrated by DD Dowden, feature the common flowers and other plants of individual states. Three to four plants are illustrated on each page. Each description includes height, other common names, flowering time, and general notes on use, source of the name, etc. Designed for children age 8 and above.

51. **Johnson Nature Series**. These books are primarily natural histories of a group of animals, but several also include an identification section in the back, such as the hummingbird and owl volumes.

52. **Kansas Nature Guides**. The University of Kansas Press has published a few guides for the state. To date, the series includes guides to mushrooms, and reptiles.

53. **Kentucky Nature Studies**. A series of about 10 volumes covering the plants and animals of Kentucky. They are basically a series of state floras and faunas written for non-scientists and illustrated with color and black-and-white photos. Not all are "true" field guides (one is a bird finding guide, for instance).

54. **Key Guide Series**. This Australian series, mostly written by Leonard Cronin and illustrated by Marion Westmacott, covers Australian mammals, wildflowers, trees, and palms. They feature a visual key to facilitate identification, maps, extensive descriptions, and natural history. A particularly attractive series, though slightly oversized.

55. **Kingfisher Field Guide**. Kingfisher is a British publisher, and many of the field guides originally published in the 1980s have been reprinted by Larousse. This series features multiple color illustrations for each species. The layout features one species per page, with sections for illustrations, illustrations of similar species, and standardized descriptions of each important part (wings of birds, sepals of plants, etc.) listed in tabular format.

56. **Lewis Clark's Field Guide Series**. Several pamphlets written by Lewis Clark covering the flowers of the Pacific Northwest. They have color photos and general notes on the region.

57. **M Caribbean Pocket Natural History Series**. This series of small, paperback tourist guides is published by Macmillan. Topics include flowers, birds, and useful plants. Most have color illustrations.

58. **Mac's Field Guide Series**. Laminated 12x7 inches (30x17cm) cards created by Craig MacGowan. They feature color illustrations of 50–100 common plants or animals of the Western United States. The only information included on each species is its size and habitat. Since they cover only a few species, they are not for the hard-core naturalist, but they are attractive and inexpensive, making them good souvenirs. They are also good candidates for carrying on a walk by the ocean.

59. **Macmillan Field Guide Series**. This less well-known series of North American guides was first published in the late 1980s. There are volumes covering birds, mushrooms, wildflowers, trees, rocks and minerals, fossils, astronomy, and the weather. They are slimmer than most of the competing guides, and have either color illustrations or color photographs. The authors are all well-known experts.

60. **Massachusetts Audubon Society's laminated cards**. This series of folded, laminated cards is not cataloged as a series, but is a fairly uniform series of guides to the flora and fauna of ecosystems in Massachusetts, such as the seashore and ponds. A similar guide that covers the cloud forests of Costa Rica is labeled as part of a *Habitat Series*, but is apparently the only one to date.

61. **Millie and Cyndi's Pocket Nature Guides**. A series of truly pocket-sized, stapled guides. They are hand-lettered and feature very attractive color illustrations and colored maps. Each includes 40–50 common species of plants, insects, or birds. Most volumes cover the western United States, though some are more comprehensive. They are almost too pretty to take into the field, but would make nice gifts or souvenirs.

> ❑ **Mini Fact Finders**. See Barron's Mini Fact Finders, entry 10.

62. **Mitchell Beazley Pocket Guides**. Although these titles are often not cataloged as a series, they form a coherent group of pocket-sized field guides with small but clear illustrations and icons showing standard information. In addition to the usual "natural" subjects, there are many volumes covering beer, wine, cheeses, etc. Many of these have been published in the U.S. by Simon and Schuster as *Simon and Schuster Pocket Guides*.

63. **National Audubon Society Field Guide Series**. The National Audubon Society publishes a number of series, all of which feature color photographs rather than paintings. The main series is flexibound (flexible plastic covers) and features an introductory chapter followed by color plates (two to three species per page) and a final species accounts section. The species accounts generally include range maps, descriptions, and brief natural history. The photographs are uniformly clear and attractive, though many purists argue photographs are not as helpful for identification as paintings are.

64. **National Audubon Society First Field Guide**. Not to be confused with the Audubon Society's Beginner's Guide series (see entry 6), these are a series of children's guides to common topics such as birds, flowers, and the weather. The standard format features an extensive introductory section, followed by descriptions and color photos of 50 common species, accompanied by another 125 species that receive less detailed coverage. Each field guide comes with a pocket-sized card showing thumbnail-sized photos of the 50 main species. First published in 1998, the plan is to publish 4 new guides per year.

65. **National Audubon Society Regional Field Guide Series**. This is a recent entry into the all-in-one market, and includes more subjects than most. The guides include the geology, weather, and star maps for each area in addition to the usual plants and animals, plus descriptions of parks and preserves. They are fairly small and feature brief descriptions and rather small, though high-quality, color photographs. Volumes planned for 1999 include guides for the Mid-Atlantic States and the Rocky Mountain States.

66. **Nature Company Guides**. The Nature Company has published several books for North America which are a cross between an attractive natural history book and an oversized field guide. They are hardbacks and far too large to be carried into the field, covering one species per page, but should prove useful for finding illustrations and basic facts about the plants and animals covered once their user returns home.

67. **Nature Finder**. A group of pocket-sized guides published by the Nature Study Guild that are clearly a series, though not identified as such. Most of the entries have "finder" in the title (*Sierra Flower Finder*, *Pacific Coast Mammal Finder*, etc.). All consist of a book-length key, with black-and-white illustrations. Most include distribution maps and natural history. The series has been published since the early 1970s, and about 22 guides are currently in print.

68. **The Nature of . . .** This series is published by Waterford Press and distributed by Falcon Press. There are about 8 titles in the series, all written by James Kavanagh and illustrated by Raymond Leung. They cover about 300 common plants and animals of a region, plus numerous natural attractions. A very attractive all-in-one guide series. Two similar titles, *Nature Alberta* and *Nature BC*, were published by Lone Pine.

69. **New Generation Guides**. Published by Collins in the U.K. and the University of Texas Press in the U.S. Subjects include butterflies, birds, wildflowers, and fungi, all of Britain and Europe. This series is intended for the naturalist who wants to go beyond mere identification. About one-half to one-third of each volume consists of a "Directory of Species" with the usual field guide-style identification information and color illustrations, followed by an extensive section on the natural history of that group of organisms.

70. **North American Wildlife**. This series of books is taken from the *Reader's Digest North American Wildlife*. The original volume was laid out very much in the manner of a field guide, but was too large to carry into the field. The repackaged volumes are smaller and feature color illustrations and range maps with brief descriptions.

71. **Northwest Naturalist Books**. Published by the University of Idaho Press, these guides cover the natural history of Idaho and the Pacific Northwest. There are several guides covering the flora and fauna of the region, including mushrooms, birds, fishes, and butterflies.

72. **The Observer's Pocket Series**. A British series of 80 or so truly pocket-sized guides published in the 1940s to 1970s by Frederick Warne and Co. Titles include the usual wild plants and animals, plus dogs, horses, British geology, aircraft, and much more. They usually have color illustrations or black-and-white photos and extensive description and natural history.

73. **Peterson Field Guide Coloring Book**. Not your usual field guide! These are coloring books that offer the budding naturalist a chance to learn about the natural world while exercising his or her artistry. Each volume includes species descriptions and natural history. The correct color of the plants or animals is illustrated in the front and back of each volume (though this seems a bit speculative in the case of the dinosaur book). Most of the "real" Peterson field guides have a coloring book equivalent.

74. **Peterson Field Guide Series**. The granddaddy of them all, the Peterson Field Guide series began with Roger Tory Peterson's *Field Guide to the Birds* and is currently up to about 51 volumes. Generally, they feature color illustrations, often separate from the text, and range maps that may or may not be integrated with the text. The "Peterson System" consists of small arrows pointing at the most important field marks, and is used in all volumes in the Peterson Field Guide series. Most of the guides are specific to one region of North America, though there are also guides to the birds and mammals of Britain and Europe. Beginning in 1998, the format of some of the volumes has been updated. While the illustrations remain in plates separate from the text, they now appear in the front of each revised volume instead of in the middle. The text of the species accounts appears in back, and small color photos of most species appears with each description.

75. **Peterson First Guides**. A pocket-sized series of beginner's guides. Most are written by the same author who wrote the corresponding Peterson Field Guide, and often include the same illustrations. They cover the most common 100 or so species.

76. **Peterson FlashGuides**. This new series consists of several laminated cards, each covering the most common birds, animals or plants of North America. Folded, they measure about 4x9 inches (12x22 cm) and expand to about 32x18 inches (70x33 cm). They feature illustrations and descriptions from the regular Peterson Field Guide Series. Examples include mushrooms, mammals, hawks, and trailside birds.

77. **Photographic Field Guide (New Holland)**. This pocket-sized series features photos, small maps, and concise descriptions with two species covered per page. There are color thumbnail tabs at the edge of each page for easy identification of families of organisms. The series is designed for travelers, covering the most common or conspicuous organisms of tourist destinations. Also published in the U.S. by Ralph Curtis Publishing.

78. **Pica Press Identification Guides**. These are very similar to the Christopher Helm Identification Guides (see entry 13). They are published in the U.K. by Pica Press and in the U.S. by Yale University Press. Pica Press is an imprint of Christopher Helm.

79. **Pictured Key Nature Series**. This series was first published in 1944 and features about 40 titles, many of them still in print. They are aimed at a student audience with extensive background information

and include some subjects which most amateur naturalists would find uninteresting (trematodes, for instance). Unlike most field guides, each volume is a book-length key, with black-and-white photos or illustrations and paragraph-length descriptions of each genus or species as it is keyed out. Some volumes include range maps.

80. **Pocket Guide** (Dragon's World). This spiralbound series is twin to the *American Nature Series* (see entry 2) in the U.S. and features similar covers and design. Some British titles are endorsed by the RSNC (Royal Society for Nature Conservation). Many volumes in the series are also published in Spain by Libros Cupula as the Pequeñas Guías de la Naturaleza series.

81. **Pocket Natural History**. A series of pocket-sized pamphlets or small books published by the Cleveland Museum of Natural History beginning in 1922. As the series statement says, "Some are field guides, others are simple stories gathered from widely scattered studies of specialties." Some volumes were co-published in the museum's botany and zoology series.

82. **The Pocket Naturalist**. A series of laminated cards featuring color illustrations of about 150 species of plants or animals of the United States. Each guide also includes a regional map showing good sites for viewing the plants or animals. There are numerous sub-series, such as the *Trees and Wildflowers* and *Wildlife* titles. They are published by Waterford Press and distributed by Falcon Press, written by James Kavanagh and illustrated by Raymond Leung. The folded size is 8½x3½ inches, opening to 8½x22 inches. The same author and illustrator created the *Nature of* . . . series (see entry 68).

83. **Princeton University Bird Taxonomic Series**. Princeton University Press is one of the North American publishers of the Christopher Helm Identification Guide Series (see entry 13). The Princeton University Press volumes are not identified as a series.

84. **Princeton University** publishes a less-uniform series of regional bird field guides, that are also not identified as a series but indisputably belong to the same family. They tend to be hefty, scholarly, comprehensive guides to the birds of a region. All feature color illustrations, some integrated with the species accounts and some in separate sections. Most of the volumes are for birds of the Americas or Australasia.

85. **Putnam's Nature Field Books**. The classic early pre-Peterson series, published between 1910–1970. There are numerous small field books in the series, covering all the usual subjects: wildflowers, mammals, ferns, mushrooms, birds, rocks, insects, etc. They feature color and black-and-white illustrations, and some include black-and-white photographs. Some also include extensive nature-study information, and all are written in a far wordier manner than modern field guides. The older volumes have a delightfully quaint air.

86. **Quick-Key Field Identification Guides**. A series of guides written in the 1960s, consisting of a pocket guide with black-and-white illustrations and a set of punched cards. Identification could be made by lining up the holes in the punched cards. Ah, technology!

87. **Rare Plant Field Guide**. The U.S. Forest Service, Bureau of Land Management, and U.S. Fish and Wildlife Service have published a number of rare plant guides that provide information on the threatened, endangered, or sensitive plants of the western states. They are in looseleaf format.

88. **Reef Watcher's**. A series of two attractive pocket-sized field guides to the fishes of Hawaii and Guam. They were written by two guides from the Blue Kirio diving company of Hawaii, and are in English and Japanese.

89. **Sea Grant Field Guide Series**. Several pamphlets published by the University of Miami's Sea Grant Program covering marine life of economic importance. They are rather technical in nature and low-budget.

90. **Seahawk Submersible Guides**. A series of 12 6x9 inches cards made of the same plastic as credit cards that illustrate about 50–75 common fishes or marine invertebrates. All are by Jerry or Idaz Greenberg. The cards are available for numerous popular diving areas of the world.

91. **Sierra Nevada Field Cards**. Another example of the laminated card featuring the most commonly seen species, these are intended for beginners and visitors to the Sierra Nevada mountains of the western U.S. They feature color illustrations of about 25 species on a 9½x6 inches (24x16 cm) cards. The only information on each species is its size and habitat. Since each card covers only a few species, they are not for the hard-core naturalist, but they are attractive and inexpensive, making them good souvenirs.

92. **Simon and Schuster's Guides**. A reprint series, translated from the Italian, originally published by Arnoldo Mondadori. Most volumes have also been published in the U.K. by Macdonald (as *Macdonald Encyclopaedias*), in France by Nathan (some with the series title Guides Nathan Nature), in Germany by BLV Verlagsgesellschaft, and in Spain by Grijalbo (some as the Guías de naturaleza series). Most of the books in the series cover plants or animals worldwide and feature color photos.

    ❏ **Simon and Schuster's Pocket Guides**. See Mitchell Beazley Pocket Guides, entry 62.

93. **South African Wild Flower Guide**. A series of 6 field guides covering the flowers of South Africa, arranged by region. Most of the volumes are up to a second edition at the time of writing. Some have color illustrations while others have color photos.

94. **Southwestern Monuments Association. Popular Series**. Several paperback field guides covering the flora and fauna of the Southwestern United States. They are paperbacks with black-and-white illustrations, and are organized by type of plant or animal (ferns, colored flowers, etc.). In many of the guides each group is printed on different colored pages.

95. **Stokes Nature Guides**. The volumes in this series are not identification guides, but rather are natural histories, covering the habits or ecology of animals species by species. They provide extensive information on, for instance, bird behavior or tracking mammals.

96. **Taylor's Guides to Gardening**. A series of colorful, field guide-sized identification guides to various types of cultivated plants, such as orchids, roses, or ornamental shrubs. Each volume is arranged with introductory chapters followed by color photos of the plants, with species or cultivar accounts at the end. Most include propagation or cultivation information for the plants covered. The "field" for this series is the botanical garden or nursery.

    ❏ **Texas Monthly Field Guide Series**. See Gulf's Field Guide Series, entry 46.

97. **Trailside Series**. "A group of guides devoted to interpreting the flora and fauna of the Pacific Northwest." These guides are published by the Seattle Audubon Society and include guides on the mammals, amphibians, reptiles, and wetland plants of the Pacific Northwest.

98. **Treasures of Nature**. This University of Texas Press had only one guide at the time of writing.

99. **Usborne Guides**. This series is more advanced than the Usborne Spotter's guides (see below). They are small and lavishly illustrated. Some illustrations in this series were also published in the Spotter's Guide Series.

100. **Usborne Spotter's Guides**. This pocket-sized series is designed for children and beginners. Each volume includes color illustrations and descriptions, with a circle next to each plant or animal name so the user can check off each species spotted. Most also include a scorecard; rarer species are given a higher score. Volumes include wild animals and plants and domesticated animals. Many volumes in the series have also been published in Spain by Juventud as part of their Guias de naturaleza series.

101. **W. L. Moody, Jr. Natural History Series**. A series of field guides covering subjects in Texas, published by Texas A & M University Press. Illustration type and subject varies; topics covered include bats, grasses, fishes, and birds of Texas. The guides generally cover natural history in more detail than most field guides.

   ❑ **Waterford Field Guides**. See entries 68 and 82.

102. **Wayside and Woodland Series**. A long-running British series published by Frederick Warne and Co. from the 1910s to 1970s. Larger than Warne's other field guide series, *The Observer's Pocket Series* (see entry 72). The Wayside and Woodland series has extensive descriptions and color illustrations and/or black-and-white photos. Covered are trees, insects, flowers, ferns, etc.

103. **Wild About Series**. A collection of three to four all-in-one guides to the flora and fauna of popular tourist areas in South Africa. Each covers about 300 species, most of them animals. They have color photos and feature color-coded pages indicating the group of plants or animals in that section.

104. **Wildlife Traveler's Companion**. A series of about four guidebooks that cover primarily European areas and are published by Crowood Press. Each volume includes a description of the country or region covered, followed by lengthy descriptions of 150 wildlife viewing sites such as parks. These two sections include color photos and maps, and are followed by a field guide section with color illustrations and descriptions of about 150 common species

of trees, flowers, butterflies, amphibians and reptiles, birds, and mammals. This series is primarily about where to go to find wild plants and animals, not how to identify them.

105. **Wildlife Watcher's Guides.** This colorful collection of wildlife guides to various national parks and vicinities includes color photos and natural history of each common or conspicuous species covered. Most have about 30 species of mammals and 10–20 birds. Good viewing sites are also included. Unlike a number of similar wildlife viewing guides, the emphasis here is on the animals, not where to find them.

Flora and Fauna

The field guides that are found in this chapter are either all-in-one guides that cover the common plants and animals of a particular area or cover both plants and animals in a particular category (poisonous organisms, garden pests). The all-in-one guides may be restricted to an ecosystem, such as the forests of western North America, or to a particular state or country, such as Florida. Often these guides are designed for travelers and cover the most common or conspicuous plants and animals of the area. These visitor's guides have been available for the United States and other popular tourist destinations for a long time. Guides to Central and South American ecotourist destinations are a more recent development. The tourist guides may also include information on the natural history, geology, and human history of the region. One problem with all-in-one guides is that they can cover only a limited number of species, so users will need other guides to identify organisms that are not included in the individual titles.

NORTH AMERICA

106. Cox, Gerald. **Wintersigns in the Snow**. Rev. ed. New York: M. Kesend, 1984. 80p. $5.95pa. ISBN 0935576118pa.

     Tracks of 40 species of animals, plus 22 trees in winter; black-and-white illustrations; pocket sized with attractive calligraphy.

107. Fitzsimons, Cecilia. **An Instant Guide to Seashore Life**. New York: Bonanza, 1989. 125p. ISBN 0517691116.

     108 species, algae to mammals; color illustrations; checklist; pocket sized. Covers primarily marine invertebrates; includes information on range, similar species, and type of beach where the shore life is found.

108. Foster, Steven, and Roger A. Caras. **A Field Guide to Venomous Animals and Poisonous Plants: North America North of Mexico**. Boston: Houghton Mifflin, 1998. 366p. (Peterson Field Guide Series, No. 46). $18.00pa. ISBN 0395515947; 039593608Xpa.

      90 animal and 250 plant species, including fungi; line drawings and color photos.

109. Kricher, John C. **Peterson First Guide to Forests**. Boston: Houghton Mifflin, 1995. 128p. $4.95pa. ISBN 0395717604pa.

      600 common species of plants and animals found in 48 North American forests. Illustrated by Gordon Morrison.

110. Kricher, John C. **Peterson First Guide to Seashores**. Boston: Houghton Mifflin, 1992. 128p. $4.95pa. ISBN 0395619017pa.

      300 species of plants and animals; color illustrations; arranged by habitat. Illustrated by Gordon Morrison.

111. Landry, Sarah. **Peterson First Guide to Urban Wildlife**. Boston: Houghton Mifflin, 1994. 128p. $5.95pa. ISBN 039593544Xpa.

      250 species commonly found near cities, including microscopic organisms, fungi, plants, and animals; color illustrations. Illustrated by the author.

112. McManners, Hugh. **The Backpacker's Handbook**. 1st American ed. New York: Dorling Kindersley; Distributed by Houghton Mifflin, 1995. 160p. $14.95. ISBN 1564588521.

      A backpack-sized guide to getting started, equipment, getting around, camping, and emergencies. Color photos and illustrations. It is intended for worldwide use, so while it identifies some edible plants and dangerous animals, the few species included are from so many areas that those sections are not useful except as a general warning. Other sections, on survival and equipment for instance, are more generally useful.

113. Mohrhardt, David, and Richard E. Schinkel. **Suburban Nature Guide: How to Discover and Identify the Wildlife in Your Backyard**. Harrisburg, PA: Stackpole Books, 1991. 252p. $16.95. ISBN 0811730808.

      Covers about 350 species of trees, shrubs, flowers, ferns, grasses, fungi, insects, spiders, amphibians, reptiles, birds, and mammals; line drawings; natural history.

114. Pielou, E. C. **A Naturalist's Guide to the Arctic**. Chicago: University of Chicago Press, 1994. 327p. $57.00; $19.95pa. ISBN 0226668134; 0226668142pa.

Extensive natural history covering atmospheric and geological phenomena, plants, animals, etc.; field guide sections identify the most common or conspicuous plants, birds, mammals, fish, and insects. Line drawings.

115.   Reid, George Kell. **Pond Life: A Guide to Common Plants and Animals of North American Ponds and Lakes**. New York: Golden, 1990. 160p. (Golden Guide). $6.00pa. ISBN 030763535X; 0307240177pa.

500 species; color illustrations. Illustrated by Sally Kaicher and Tom Dolan.

116.   Rhodes, Frank, Harold Trevor, Herbert S. Zim, and Paul R. Shaffer. **Fossils: A Guide to Prehistoric Life**. New York: Golden, 1990. 160p. (Golden Guide). $6.00pa. ISBN 0307244113pa.

500 extinct species; color illustrations; maps; geological time-scale and other background information. Illustrated by Raymond Perlman.

117.   Wernert, Susan J., ed. **Reader's Digest North American Wildlife**. Updated ed. Pleasantville, NY: Reader's Digest Association, 1998. 559p. $28.95. ISBN 0762100206.

2,000 species of mushrooms, nonflowering plants, trees, flowers, invertebrates, fish, reptiles, amphibians, birds, and mammals; brief visual keys; color illustrations; maps. Large for a field guide, but covers all major groups of living organisms.

118.   Winchester, A. M., and H. E. Jaques. **How to Know the Living Things**. 2nd ed. Dubuque, IA: W. C. Brown, 1981. 173p. (Pictured Key Nature Series). $25.80pa. ISBN 0697047806pa; 0697047784.

Dichotomous key throughout; line drawings; includes background information on classification and nature study. Keys out all groups of organisms, from viruses to mammals, to the level of the order (e.g., rodents or conifers). Representative species of each order are illustrated. Revised edition of *Living Things, How to Know Them* by H. E. Jaques.

119.   Zim, Herbert Spencer, and Lester Ingle. **Seashores: A Guide to Animals and Plants along the Beaches**. New York: Golden, 1991. 160p. (Golden Guide). $5.95pa. ISBN 0307644960; 0307244962pa.

475 subjects are covered, including algae, flowering plants found along the seashore, seashells and other invertebrates, and shorebirds; color illustrations. Illustrated by Dot and Sy Barlowe.

Eastern North America

120. Alden, Peter, et al. **National Audubon Society Field Guide to Florida**. New York: Knopf, 1998. 447p. (National Audubon Society Regional Field Guide Series). $19.95pa. ISBN 067944677Xpa.

    1,000 species of mushrooms, non-flowering plants, trees, flowers, invertebrates, amphibians, reptiles, birds, and mammals; color photos. Also includes 16 star charts, plus information on geology, fossils, habitats, weather, and parks and preserves of Florida.

121. Alden, Peter, et al. **National Audubon Society Field Guide to New England**. New York: Knopf, 1998. 447p. (National Audubon Society Regional Field Guide Series). $19.95pa. ISBN 0679446761pa.

    1,000 species of mushrooms, non-flowering plants, trees, flowers, invertebrates, amphibians, reptiles, birds, and mammals; color photos. Also includes 16 star charts, plus information on geology, fossils, habitats, weather, and parks and preserves of the New England region.

122. Amos, William Hopkins, and Stephen H. Amos. **Atlantic and Gulf Coasts**. New York: Knopf, 1985. 670p. (Audubon Society Nature Guides). $19.00 (flexi). ISBN 0394731093 (flexi).

    600 common species of birds, plants, seashore creatures, fish, and whales; color photos, line drawings; maps.

123. Beatte, Brian, and Brett Dufur. **River Valley Companion: A Nature Guide**. Columbia, MO: Pebble, 1997. 247p. $14.95pa. ISBN 0964662515pa.

    400 species of plants and animals, plus rocks, geological features, and the weather; black-and-white illustration; life list. Covers the natural history of the Missouri River Valley and the Katy Trail of Missouri. Illustrated by Maggie Riesenmy.

124. Benyus, Janine M. **Northwoods Wildlife: A Watcher's Guide to Habitats**. Minocqua, WI: NorthWord, 1989. 453p. ISBN 1559710039.

    300 species; line drawings; arranged by habitat; natural history. Also available in an abridged Knapsack edition (ISBN 1559711361).

125. Boyd, Howard P. **A Field Guide to the Pine Barrens of New Jersey: Its Flora, Fauna, Ecology, and Historic Sites**. Medford, NJ: Plexus, 1991. 423p. $32.95; $22.95pa. ISBN 0937548189; 0937548197pa.

    360 species of plants and 450 species of insects are covered, plus fish, amphibians, reptiles, birds, and mammals; black-and-white

illustrations; extensive introduction and natural history. Illustrated by Mary Pat Finelli.

126. Brown, Lauren. **Grasslands**. New York: Knopf, 1985. 606p. (Audubon Society Nature Guides). $19.95pa. ISBN 0394731212pa.
      600 common species of birds, wildflowers, trees, grasses, and insects; color photos and line drawings; maps.

127. Colin, Patrick L. **Caribbean Reef Invertebrates and Plants: A Field Guide to the Invertebrates and Plants Occurring on Coral Reefs of the Caribbean, the Bahamas and Florida**. Neptune City, NJ: T. F. H., 1978. 512p. ISBN 0876664605.
      Few plants; color photos; rather heavy for a field guide.

128. Conrader, Jay M. **The Northwoods Wildlife Region**. Happy Camp, CA: Naturegraph, 1984. 188p. ISBN 087961126X; 0879611278pa.
      250 species of plants and animals; color illustrations. Covers the Great Lakes region and southern Canada. Illustrated by Constance Conrader.

129. Daniel, Glenda. **Dune Country: A Hiker's Guide to the Indiana Dunes**. Rev. ed. Athens, OH: Swallow, 1984. 176p. $12.95pa. ISBN 080400854Xpa.
      175 species of plants and 80 species of animals are covered; black-and-white illustrations; primarily a natural history covering the geology, natural history, and trails of Indiana Dunes National Seashore. Illustrated by Carol Lerner.

130. Day, Cherie H. **Life on Intertidal Rocks: A Guide to Marine Life of the North Atlantic Coast**. Berkeley, CA: Nature Study Guild, 1987. 61p. $3.00pa. ISBN 0912550155pa.
      100 species of plants and animals; black-and-white illustrations; pocket sized.

131. Frost, Ed. **Discovering Nature in New England's Mountains**. Portsmouth, NH: Glove Compartment Books, 1993. 112p. $9.95pa. ISBN 0961880643pa.
      60 species of trees, flowers, and animals; black-and-white illustrations; natural history. Illustrated by Susan K. Hamilton.

132. Gosner, Kenneth L. **A Field Guide to the Atlantic Seashore**. Boston: Houghton Mifflin, 1979. 329p. (Peterson Field Guide Series, No. 24). $24.95; $17.95pa. ISBN 0395243793; 0395318289pa.
      140 species of seaweeds and 907 species of invertebrates; black-and-white and color illustrations. Illustrated by the author.

133. Kaplan, Eugene H. **A Field Guide to Southeastern and Caribbean Seashores: Cape Hatteras to the Gulf Coast, Florida, and the Caribbean**. Boston: Houghton Mifflin, 1988. 425p. (Peterson Field Guide Series, No. 36). $22.45; $18.00pa. ISBN 039531321X; 0395468116pa.

　　1,000 species of plants, algae, and invertebrates; black-and-white and color illustrations. Illustrated by Susan L. Kaplan.

134. Kavanagh, James. **The Nature of Florida: An Introduction to Familiar Plants and Animals and Natural Attractions**. Blaine, WA: Waterford, 1997. 175p. (Waterford Field Guides). $17.95pa. ISBN 0964022575pa.

　　310 species; color illustrations; checklists; 5 natural areas. Illustrated by Raymond Leung.

135. Kricher, John C. **A Field Guide to Eastern Forests of North America**. Boston: Houghton Mifflin, 1998. (Peterson Field Guide Series, No. 37). $19.00pa. ISBN 0395928958pa.

　　530 species of plants and animals; color illustrations of organisms and color photos of habitats; arranged by habitat. Emphasis is more on natural history than on identification. Revised edition of *A Field Guide to the Ecology of Eastern Forests, North America*. Illustrated by Gordon Morrison.

136. Mader, Sylvia S. **Martha's Vineyard Nature Guide**. Edgartown, MA: Mader Enterprises, 1985. 96p. $7.95pa. ISBN 03140346Xpa.

　　106 species; color photos and illustrations; arranged by habitat. Illustrated by Kathleen Hagelston.

137. Nellis, David W. **Poisonous Plants and Animals of Florida and the Caribbean**. Sarasota, FL: Pineapple, 1996. 315p. $29.95; $21.95pa. ISBN 1561641111; 1561641138pa.

　　100 species of plants and 44 species of animals are covered; color photos. Part one consists of identification, with distribution, description, and habitat for all species, plus reproduction, food, and longevity for the animals. Part 2 details the toxins and the portion of the plant or animal that produces them and describes treatments and beneficial uses.

138. Niering, William A. **Wetlands**. New York: Knopf, 1985. 638p. (Audubon Society Nature Guides). $19.95pa. ISBN 0394731476pa.

　　600 common species of trees, wildflowers, fish, insects, and birds; color photos and line drawings; maps.

139. Roberts, Mervin F. **The Tidemarsh Guide**. New York: E. P. Dutton, 1979. 240p. $5.95pa. ISBN 0933614195pa; 0525930809.

    400 species; black-and-white illustrations; all types of organisms except flowers. Casual tone; natural history. Illustrated by Sandra G. Power.

140. Sharpe, Grant William. **A Guide to Acadia National Park and the Nearby Coast of Maine**. New York: Golden, 1968. 80p. (A Golden Regional Guide).

    80 species of plants, 30 species of mammals, 40 birds, 10 fish, 7 amphibians, 7 reptiles, and 30 marine invertebrates; includes general background information about the region. Illustrated by Jane Ingraham Rupp.

141. Slack, Nancy G., and Allison W. Bell. **Field Guide to the New England Alpine Summits**. Boston: Appalachian Mountain Club, 1995. 96p. $12.95. ISBN 0878239384.

    195 species of plants and animals; color photos; natural history.

142. Steele, Frederic L. **At Timberline: A Nature Guide to the Mountains of the Northeast**. Boston: Appalachian Mountain Club, 1982. 285p. $13.95pa. ISBN 091014639Xpa.

    99 species of trees, 34 ferns, 174 flowers, 19 amphibians and reptiles, 91 birds, 29 mammals, and 17 geological formations; black-and-white illustrations, color illustrations of birds.

143. Stokes, Donald W. **A Guide to Nature in Winter: Northeast and North Central North America**. Boston: Little, Brown, 1976. 374p. $18.95; $14.95pa. ISBN 0316817201; 0316817236pa.

    Consists of a number of field guides, each with general information, a visual key, and description of species. Guides include winter weeds, snow, trees, insects, birds and abandoned nests, mushrooms, tracks, and evergreens. Black-and-white illustrations. Illustrated by Deborah Princer and the author.

144. Sutton, Ann, and Myron Sutton. **Eastern Forests**. New York: Knopf, 1985. 638p. (Audubon Society Nature Guides). $19.95pa. ISBN 0394731263pa.

    600 common species found in eastern woodlands; color photos and line drawings; maps.

145. White, Christopher P. **Chesapeake Bay: Nature of the Estuary: A Field Guide**. Centreville, MD: Tidewater Publishers, 1989. 212p. $14.95pa. ISBN 0870333518pa.

500 species of plants and animals; black-and-white illustrations; arranged by habitat; natural history. Illustrated by Karen Teramura.

146. Zim, Herbert Spencer. **A Guide to Everglades National Park and the Nearby Florida Keys**. New York: Golden, 1983. Updated ed. 80p. (A Golden Guide). $3.95pa. ISBN 0307240053pa.

Covers 12 mammals, 35 birds, 33 amphibians and reptiles; 17 fish, 20 invertebrates, and 58 plants; color illustration; natural history. Illustrated by Russ Smiley.

Western North America

147. Alden, Peter, et al. **National Audubon Society Field Guide to California**. New York: Knopf, 1998. 447p. (National Audubon Society Regional Field Guide Series). $19.95pa. ISBN 0679446788pa.

1,000 species of mushrooms, non-flowering plants, trees, flowers, invertebrates, amphibians, reptiles, birds, and mammals; color photos. Also includes 16 star charts, plus information on geology, fossils, habitats, weather, and parks and preserves of California.

148. Alden, Peter, et al. **National Audubon Society Field Guide to the Pacific Northwest**. New York: Knopf, 1998. 448p. (National Audubon Society Regional Field Guide Series). $19.95pa. ISBN 0679446796pa.

1,000 species of mushrooms, non-flowering plants, trees, flowers, invertebrates, amphibians, reptiles, birds, and mammals; color photos. Also includes 16 star charts, plus information on geology, fossils, habitats, weather, and parks and preserves of the Pacific Northwest.

149. Fielding, Ann, and Ed Robinson. **An Underwater Guide to Hawai'i**. Honolulu: University of Hawaii Press, 1987. 156p. $19.95. ISBN 0824811046.

200 plants and animals; color photos; half of book consists of background and natural history.

150. Hall, Clarence A. **Natural History of the White—Inyo Range, Eastern California**. Berkeley, CA: University of California Press, 1991. 536p. (California Natural History Guides, No. 55). ISBN 0520068955; 0520068963pa.

340 species or groups of flowers, 16 trees, 113 insects, 4 fish, 8 amphibians, 37 reptiles, 85 birds, and 45 mammals; tree key, wildflower key by color and form; line drawings, color photos

separate from text. This is a natural history guide, covering the weather, geology, and archaeology of the area as well as common plant and animal species. A handy guide for residents of the area to keep on hand, but a rather hefty field companion.

151.   Harbo, Rick M. **The Edible Seashore: Pacific Shores Cookbook and Guide**. Surrey, B.C.: Hancock House, 1988. 62p. $5.95pa. ISBN 0888391994pa.

3 seaweed species, invertebrates, and 14 animal species; color photos; includes recipes and information on catching, storing, and serving.

152.   Hedgpeth, Joel Walker. **Introduction to Seashore Life of the San Francisco Bay Region and the Coast of Northern California**. Berkeley, CA: University of California Press, 1962. 136p. (California Natural History Guides, No. 9). $11.95pa. ISBN 0520029925; 0520005449pa.

Keys; line drawings; no fishes, one page of mammals; brief descriptions by families. Illustrated by the author and Lynn Rudy.

153.   Kavanagh, James. **Nature Alberta: An Illustrated Guide to Common Plants and Animals**. Edmonton, Alta.: Lone Pine, 1991. 191p. $11.95pa. ISBN 091943391Xpa.

351 species of trees, plants, fish, amphibians, birds, and mammals; color illustrations; maps; checklist. Includes list of wildlife sanctuaries and provincial parks.

154.   Kavanagh, James. **Nature BC: An Illustrated Guide to Common Plants and Animals**. Edmonton, Alta.: Lone Pine, 1993. 173p. $14.95pa. ISBN 1551050366pa.

50 mammals, 80 birds, 8 reptiles, 8 amphibians, 24 fishes, 50 trees and shrubs, 70 flowers; color illustrations; maps; checklist. Also includes wildlife viewing areas and sanctuaries.

155.   Kavanagh, James. **The Nature of Alaska: An Introduction to Familiar Plants and Animals and Natural Attractions**. Blaine, WA: Waterford, 1997. 175p. (Waterford Field Guides). $17.95pa. ISBN 0964022559pa.

275 species of plants and animals; color illustrations; checklists. Illustrated by Raymond Leung.

156.   Kavanagh, James. **The Nature of Arizona: An Introduction to Familiar Plants and Animals and Natural Attractions**. San Francisco, CA: Waterford, 1996. 175p. (Waterford Field Guides). $17.95pa. ISBN 0964022583pa.

275 species of flowers, trees, invertebrates, fishes, reptiles and amphibians, birds, and mammals; color illustrations; check-list. Also includes descriptions of 100 natural attractions. Illus-trated by Raymond Leung.

157. Kavanagh, James. **The Nature of California: An Introduction to Familiar Plants and Animals and Natural Attractions**. 2nd ed. San Francisco, CA: Waterford, 1997. 175p. (Waterford Field Guides). $17.95pa. ISBN 0964022540pa.

    325 species of plants and animals and 85 natural attractions; color illustrations; checklist. Descriptions include size, habitat, and general comments. Illustrated by Raymond Leung.

158. Kavanagh, James. **The Nature of Colorado: An Introduction to Familiar Plants and Animals and Natural Attractions**. San Francisco, CA: Waterford, 1997. 175p. (Waterford Field Guides). $17.95pa. ISBN 0964022567pa.

    Not seen; see series entry 68. Illustrated by Raymond Leung.

159. Kavanagh, James. **The Nature of Washington: An Introduction to Familiar Plants and Animals and Natural Attractions**. San Francisco, CA: Waterford, 1997. 175p. (Waterford Field Guides). $17.95pa. ISBN 0964022532pa.

    Not seen; see series entry 68. Illustrated by Raymond Leung.

160. Kozloff, Eugene N. **Seashore Life of the Northern Pacific Coast: An Illustrated Guide to Northern California, Oregon, Washington, and British Columbia**. Seattle, WA: University of Washington Press, 1983. 370p. $31.95pa. ISBN 029560302; 0295960302pa.

    650 species; black-and-white photos and line drawings, some color photos; arranged by habitat; natural history. Oversized. Revised edition of *Seashore Life of Puget Sound, the Strait of Georgia, and the San Juan Archipelago*.

161. Kozloff, Eugene N. **Plants and Animals of the Pacific Northwest: An Illustrated Guide to the Natural History of Western Oregon, Washington, and British Columbia**. Seattle, WA: University of Washington Press, 1978. 264p. $29.95pa. ISBN 029595597Xpa.

    560 species; color photos; by habitat. This is more a natural history than a field guide.

162. Kricher, John C. **A Field Guide to California and Pacific North-west Forests**. Boston: Houghton Mifflin, 1998. 480p. (Peterson Field Guide Series, No. 50). $19.00pa. ISBN 0395928966pa.

530 species of plants and animals; color illustrations of organisms and color photos of habitats; arranged by habitat. Includes forests of Canada and Alaska. Revised edition of *A Field Guide to the Ecology of Western Forests*. Illustrated by Gordon Morrison.

163. Kricher, John C. **A Field Guide to Rocky Mountain and Southwest Forests**. Boston: Houghton Mifflin, 1998. 480p. (Peterson Field Guide Series, No. 51). $19.00pa. ISBN 0395928974pa.

530 species of plants and animals; color illustrations of organisms and color photos of habitats; arranged by habitat. Includes forests of Canada and Alaska. Revised edition of *A Field Guide to the Ecology of Western Forests*. Illustrated by Gordon Morrison.

164. MacMahon, James A. **Deserts**. A Chanticleer Press ed. New York: Knopf, 1985. 638p. (Audubon Society Nature Guides). $19.95pa. ISBN 0394731395pa.

600 common species of wildflowers, cacti, trees, birds, reptiles, insects, etc.; color photos and line drawings; maps.

165. Mathews, Daniel. **Cascade—Olympic Natural History**. Portland, OR: Raven Editions in conjunction with the Audubon Society of Portland, 1988. 625p. $22.50pa. ISBN 0962078204pa.

700 species of plants and animals; line drawings and color photos; includes rock types, geology, and climate of the area.

166. McConnaughey, Bayard Harlow, and Evelyn McConnaughey. **Pacific Coast**. New York: Knopf, 1985. 633p. (Audubon Society Nature Guides). $19.95pa. ISBN 0394731301pa.

600 species (170 seashells, 40 marine mammals, 90 fish, 160 marine invertebrates, 25 seaweeds, 90 birds, 15 land mammals); color photos.

167. Miller, Millie. **Desert Critters: Plants and Animals of the Southwest**. Boulder, CO: Johnson Books, 1996. 63p. (Millie & Cyndi's Pocket Nature Guides). $5.95pa. ISBN 1555661726pa.

275 species of plants and animals; color illustrations; tracks of 13 mammals; very brief descriptions and thumbnail-sized illustrations. Pocket-sized pamphlet.

168. North, Wheeler J. **Underwater California**. Berkeley, CA: University of California Press, 1976. 276p. (California Natural History Guides, No. 39). $12.95pa. ISBN 0520030257; 0520030397pa.

700 species, algae to mammals; line drawings and color photos; natural history, including chapter on underwater photography. Illustrated by Laurence G. Jones.

169. Russo, Ron. **Hawaiian Reefs: A Natural History Guide**. San Leandro, CA: Wavecrest, 1994. 174p. $17.95pa. ISBN 0963569600pa.

    168 species; color photos; background information on reefs and their biology.

170. Smith, Kathleen M., Nancy J. Anderson, and Katherine I. Beamish, eds. **Nature West Coast: As Seen in Lighthouse Park**. Victoria, B.C.: Sono Nis, 1988. 283p. ISBN 0919203833pa.

    200 species of plants (including mosses and lichens), 25 mushrooms, 25 insects, 9 amphibians and reptiles, 65 birds, 16 mammals, and 45 marine plants and animals; black-and-white illustrations; background information. Covers the lower mainland of British Columbia, the Gulf Islands, and southeastern Vancouver Island in Canada.

171. Snively, Gloria. **Exploring the Seashore in British Columbia, Washington, and Oregon: A Guide to Shorebirds and Inter-tidal Plants and Animals**. Mercer Island, WA: Writing Works, 1978. 240p. $19.95. ISBN 0916076245.

    350 species; black-and-white illustrations; natural history. Arranged by zones. Illustrated by Philip Croft and Mark Wynja.

172. Storer, Tracy I., and Robert L. Usinger. **Sierra Nevada Natural History: An Illustrated Handbook**. Berkeley, CA: University of California Press, 1963. 374p. $15.95pa. ISBN 0520012275pa.

    750 species or groups of plants and animals; black-and-white illustrations, some color photos; includes chapters on climate, geology, and ecology of the region. Also has numerous tables to aid in identification (flowers by color, birds by habitat, etc.).

173. Whitney, Stephen. **Field Guide to the West Coast Mountains**. Vancouver, B.C.: Douglas and McIntyre, 1983. 288p. ISBN 0888944039pa.

    600 species of plants and animals; color and black-and-white illustrations. Covers northern California to southern British Columbia.

174. Whitney, Stephen. **A Field Guide to the Cascades and Olympics**. Seattle, WA: Mountaineers, 1983. 288p. $18.95pa. ISBN 0898860776pa.

600 species of ferns, wildflowers, shrubs, trees, butterflies, fish, amphibians, reptiles, birds, and mammals; color and black-and-white illustrations. Illustrated by the author.

175. Whitney, Stephen. **A Field Guide to the Grand Canyon**. 2nd ed. Seattle, WA: Mountaineers, 1996. 269p. $19.95pa. ISBN 0898864895pa.

   480 species of plants and animals; color and black-and-white illustrations, some color photos; flowers arranged by color. Guide to almost everything about the Grand Canyon, including geology, plants, and animals. Illustrated by the author.

176. Whitney, Stephen. **Western Forests**. New York: Knopf, 1985. 671p. (Audubon Society Nature Guides). $19.95pa. ISBN 0394731271pa.

   600 common species of plants and animals found in forests west of the Rockies; color photos, some line drawings; maps.

Plants

This chapter covers comprehensive plant field guides and guides to flowering plants such as weeds, cacti, and grasses that do not fit in the other plant chapters. There are many field guides that cover a variety of flowering plants such as trees, grasses, and wildflowers plus a few select non-flowering plants such as ferns or mosses. These general plant guides generally cover many more species than the flora and fauna guides discussed in Chapter 2, but suffer from the same problem of covering a limited number of the species a user might want to identify.

A comprehensive field guide to plants covers many more species than a comprehensive animal guide. For instance, a comprehensive guide to the plants of any one state would probably cover over 1,000 species, compared to approximately 600 species of birds found in all of North America. This means that most comprehensive plant field guides include little information on each species. This leaves a niche for more specialized guides that cover a smaller section of the plant world.

Most of the guides annotated in this chapter cover wild, as opposed to cultivated, plants. Some guides to cultivated plants are included, if they follow the general field guide format, which includes many illustrations and little text. There are also several weed guides, some of which include information on weed control. Because weed control methods change rapidly, most guides to weeds do not include information on control.

For other plant guides, see Chapter 2, "Flora and Fauna."

NORTH AMERICA

177. Angell, Madeline. **A Field Guide to Berries and Berrylike Fruits**. Indianapolis: Bobbs-Merrill, 1981. 250p. ISBN 067252676X; 0672526956pa.

     200 species of flowers, shrubs, trees; line drawings; arranged by color of berry at maturity. Illustrated by Marie Suilmann.

178. Baumgardt, John Philip. **How to Identify Flowering Plant Families: A Practical Guide for Horticulturists and Plant Lovers**. Portland, OR: Timber, 1982. 269p. $22.95pa. ISBN 0917304217pa.

     110 families, with emphasis on cultivated and North American plants; key; black-and-white illustrations. Each family is described, with common or interesting genera mentioned and one or two species illustrated.

179. Crockett, Lawrence J. **Wildly Successful Plants: A Handbook of North American Weeds**. New York: Macmillan, 1977. 268p. ISBN 0025288504, 0020626002pa.

     100 common weeds; line drawings; natural history. Illustrated by Joanne Bradley.

180. Cronquist, Arthur. **How to Know the Seed Plants**. Dubuque, IA: W. C. Brown, 1979. 153p. (Pictured Key Nature Series). $25.80pa. ISBN 0697047601; 069704761Xpa.

     Key to family level; black-and-white line drawings; emphasis on U.S. plants.

181. Cullen, J. **The Identification of Flowering Plant Families, Including a Key to Those Native and Cultivated in North Temperate Regions**. 4th ed. New York: Cambridge University Press, 1997. 215p. $59.95. ISBN 052158485X; 0521585503pa.

     Key to 286 families of plants, both cultivated and wild; a few black-and-white illustrations; family accounts; covers the Northern Hemisphere. This guide uses technical language, but is suitable for students or botanists, but small enough to be taken into the field. New edition of P. H. Davis and J. Cullen's 1987 volume.

182. Dawson, Elmer Yale. **How to Know the Cacti: Pictured Keys for Determining the Native Cacti of the United States and Many of the Introduced Species**. Dubuque, IA: W. C. Brown Co, 1963. 158p.

     170 species; key; black-and-white photos and illustrations.

183. Dickinson, Richard, and France Royer. **Weeds of Canada and the Northern United States: A Guide for Identification**. Edmonton: University of Alberta Press, 1998. 400p. ISBN 088864311Xpa.

150 species are covered in detail, another 100 are discussed; color photos. Also includes information on weed legislation by state or province.

184. Embertson, Jane. **Pods: Wildflowers and Weeds in Their Final Beauty**. New York: Scribner, 1979. 186p. $18.00. ISBN 0684155427.

150 species; color photos illustrating flower, pod, and dried arrangements.

185. Gunn, Charles R., and John V. Dennis. **World Guide to Tropical Drift Seeds and Fruits**. New York: Quadrangle/New York Times Book Co., 1976. 240p. ISBN 0812906160.

150 genera of seeds found drifting on beaches; key; black-and-white illustrations and photos; natural history. Illustrated by Pamela J. Paradine.

186. Heil, Kenneth D. **Familiar Cacti**. New York: Knopf, 1992. 191p. (Audubon Society Pocket Guides). $8.00pa. ISBN 0679741496pa.

80 species; color photos.

187. Jaques, H. E. **Plant Families, How to Know Them: Pictured-Keys for Determining the Families of Nearly All of the Members of the Entire Plant Kingdom**. 2nd ed. Dubuque, IA: W. C. Brown, 1949. 177p. (Pictured Key Nature Series).

Comprehensive, covering almost all plant families world-wide; key throughout; black-and-white illustrations.

188. Kapp, Ronald O. **How to Know Pollen and Spores**. Dubuque, IA: W. C. Brown, 1969. 249p. ISBN 697048497; 697048489pa. (Pictured Key Nature Series).

Keys to family or genus; black-and-white illustrations. Requires microscope.

189. Knobel, Edward. **Field Guide to the Grasses, Sedges and Rushes of the United States**. New York: Dover, 1977. 83p. $3.95pa. ISBN 048623505Xpa.

370 species; key; line drawings. Emphasis is on wild grasses, but some cultivated species are also included.

190. Lotschert, Wilhelm, and Gerhard Beese. **Collins Guide to Tropical Plants**. New York: Viking Penguin, 1989. 256p. $24.95. ISBN 0685445518.

Covers 323 species of ornamental and economically important plants; color photos; includes description, distribution, and flowering time. Translation of *Pflanzen der Tropen*. English version first published by Collins in the U.K.

191. Martin, Alexander Campbell. **Weeds**. New York: Golden, 1987. 160p. (Golden Guide). $6.00pa. ISBN 0307243532pa.
     150 species; color illustrations; maps. Illustrated by Jean D. Zallinger.

192. Montgomery, Frederick Howard. **Plants from Sea to Sea**. Toronto: Ryerson, 1966. 453p.
     1,500 species of Canadian plants; key; line drawings, some color photos.

193. Morton, Julia F. **Exotic Plants**. New York: Golden, 1971. 160p. (A Golden Nature Guide). ISBN 0307243433; 0307244164pa.
     400 tropical and sub-tropical plants; color illustrations; includes propagation information. Illustrated by Richard E. Younger.

194. Perry, Frances, and Roy Hay. **A Field Guide to Tropical and Subtropical Plants**. New York: Van Nostrand Reinhold, 1982. 136p. ISBN 0442268610; 0442268599pa.
     200 species; color illustrations.

195. Pizzetti, Mariella. **Simon and Schuster's Guide to Cacti and Succulents**. New York: Simon and Schuster, 1985. 384p. $13.99pa. ISBN 0671602314pa.
     300 species; color photos; alphabetical by genus; includes information on cultivation.

196. Pohl, Richard W. **How to Know the Grasses**. 3rd ed. Dubuque, IA: W. C. Brown, 1978. 200p. (Pictured Key Nature Series). $25.80pa. ISBN 0697048772; 0697048764pa.
     324 species; key; line drawings; maps; background information.

197. Reader's Digest Editors. **North American Wildlife: Trees and Nonflowering Plants**. New York: Reader's Digest, 1998. 269p. $17.95pa. ISBN 0762100370pa.
     500 species; half are non-flowering plants; color illustrations; maps.

198. Rosene, Walter. **A Guide to and Culture of Flowering Plants and Their Seed Important to Bobwhite Quail**. Augusta, GA: Morris Communications Corp., 1988. 170p. $26.95pa. ISBN 0961827025; 0961827017pa.

    150 plants; color photos of plants and seeds; description includes distribution and culture.

Eastern North America

199. Ahmadjian, Vernon. **Flowering Plants of Massachusetts**. Amherst: University of Massachusetts Press, 1979. 582p. $45.00. ISBN 0370232657.

    495 species described, 277 illustrated with black-and-white illustrations; key to families. Illustrated by Barry Moser.

200. Barkley, T. M. **Field Guide to the Common Weeds of Kansas**. Lawrence: University Press of Kansas, 1983. 164p. $19.95; $7.95pa. ISBN 070060233X; 0700602240pa.

    200 species; black-and-white illustrations.

201. Batson, Wade T. **A Guide to the Genera of the Plants of Eastern North America**. 4th ed. Columbia, SC: University of South Carolina Press, 1997. 203p. ISBN 0872494519pa; 1570031428.

    Comprehensive; key only; line drawings.

202. Brown, Lauren. **Grasses: An Identification Guide**. Boston: Houghton Mifflin, 1992. 240p. (Peterson Nature Library). $13.00pa. ISBN 0395628814pa.

    Covers about 150 common species, including sedges and rushes; black-and-white illustrations; key. Covers northeastern U.S. Reprint of the 1979 edition. Illustrated by the author.

203. Brown, Lauren. **Wildflowers and Winter Weeds**. New York: Norton, 1997. 252p. $13.00pa. ISBN 0393316785pa.

    225 species; key; line drawings show dried plants and seed heads; northeastern U.S. Originally published in 1986 as *Weeds in Winter*. Illustrated by the author.

204. Chambers, Brenda, et al. **Forest Plants of Central Ontario**. Redmond, WA: Lone Pine, 1996. 448p. $19.95pa. ISBN 1551050617pa.

    700 species of trees, shrubs, wildflowers, grasses, ferns, mosses, and lichens; color photos and line drawings. Descriptions include habitat and use.

205. Duncan, Wilbur Howard, and Marion B. Duncan. **The Smithsonian Guide to Seaside Plants of the Gulf and Atlantic Coasts from Louisiana to Massachusetts, Exclusive of Lower Peninsular Florida**. Washington, DC: Smithsonian Institution Press, 1987. 409p. $29.95pa. ISBN 0874743869; 0874743877pa.

    943 species; key; line drawings, color photos separate from text; flowering plants only.

206. Edsall, Marian S. **Roadside Plants and Flowers: A Traveler's Guide to the Midwest and Great Lakes Area: With a Few Familiar Off-Road Wildflowers**. Madison, WI: University of Wisconsin Press, 1985. 143p. $14.95pa. ISBN 0299097005; 0299097048pa.

    150 roadside plants and 31 off-road plants; color photos; arranged by color.

207. Eleuterius, Lionel N. **Tidal Marsh Plants**. Gretna, LA: Pelican, 1990. 168p. $27.50. ISBN 0882897950.

    200 species; line drawings, some color photos. Covers coastal salt marshes of the southern Atlantic Coast and the Gulf of Mexico. Oversized; intended for use in botany courses.

208. Evans, Douglas H. **Cactuses of Big Bend National Park**. Austin, TX: University of Texas Press, 1998. 96p. (Corrie Herring Hooks Series, No. 38). $25.00; $12.95pa. ISBN 029272098X; 0292720998pa.

    12 types of cacti; color photos.

209. Everitt, J. H., and D. Lynn Drawe. **Trees, Shrubs and Cacti of South Texas**. Lubbock, TX: Texas Tech University Press, 1993. 213p. $18.95pa. ISBN 0896722538pa; ISBN 089672252X.

    190 species; color photos. Covers the 14 southernmost counties of Texas.

210. Fassett, Norman C., and Olive S. Thomson. **Spring Flora of Wisconsin: A Manual of Plants Growing Without Cultivation and Flowering Before June 15**. 4th ed., rev. and enl. Madison: University of Wisconsin Press, 1976. 413p. $18.50; $9.95pa. ISBN 0299067505; 0299067548pa.

    Comprehensive; key; black-and-white illustrations; arranged systematically. Illustrated by Elizabeth Hollister Zimmerman and Roderic L. Thomson.

211. Gould, Frank W. **Common Texas Grasses: An Illustrated Guide**. College Station: Texas A&M University Press, 1978. 267p. (W. L. Moody, Jr., Natural History Series, No. 3). $15.95pa. ISBN 0890960577; 0890960585pa.

     150 species; key; line drawings; alphabetical by genus.

212. Grimm, William Carey. **The Illustrated Book of Wildflowers and Shrubs: The Comprehensive Field Guide to More Than 1,300 Plants of Eastern North America**. Rev. and updated. Harrisburg, PA: Stackpole, 1992. 637p. $24.95pa. ISBN 0811730859pa.

     1,300 species; keys; black-and-white illustrations; oversized.

213. Hall, David W. **Illustrated Plants of Florida and the Coastal Plain**. Gainesville, FL: Maupin House, 1993. 431p. ISBN 029895401.

     1,200 species; black-and-white illustrations; alphabetical by family. Illustrated by Edward H. Stehman.

214. Haragan, Patricia Dalton. **Weeds of Kentucky and Adjacent States: A Field Guide**. Lexington, KY: University Press of Kentucky, 1991. 278p. $29.00. ISBN 0813117437.

     160 species; black-and-white illustrations; arranged by color of flower; no information on weed control.

215. Haukos, David A., and Loren M. Smith. **Common Flora of the Playa Lakes**. Lubbock: Texas Tech University Press, 1997. 196p. $18.95pa. ISBN 0896723887pa.

     150 species; color photos; checklist; descriptions include flowering period, habitat, and abundance. Covers the plants of the temporary lakes of the Llana Estacado in Texas.

216. Hoagman, Walter J. **Great Lakes Coastal Plants: A Field Guide**. East Lansing, MI: Michigan State University, 1994. 135p. $7.95pa. ISBN 1565250087pa.

     113 species; line drawings.

217. Hoyer, Mark V., ed. **Florida Freshwater Plants: A Handbook of Common Aquatic Plants in Florida Lakes**. Gainesville, FL: University of Florida, Institute of Food and Agricultural Sciences, 1996. 264p. [no ISBN]

     100 species; color photos; maps. Includes technical information on alkalinity, mineral, and other requirements of each species.

218. Johnson, Karen L. **Wildflowers of Churchill and the Hudson Bay Region**. Winnipeg: Manitoba Museum of Man and Nature, 1987. 400p. ISBN 0920704158.

200 species; key; color photos and illustrations, line drawings; natural history. Covers trees, shrubs, and wildflowers.

219. Kraus, E. Jean Wilson. **A Guide to Ocean Dune Plants Common to North Carolina**. Chapel Hill: Published for the University of North Carolina Sea Grant College Program by the University of North Carolina Press, 1988. 72p. $8.95pa. ISBN 0807842125pa.
     50 species; simple keys; line drawings. Illustrated by the author.

220. Lamoureaux, Gisele, ed. **Plantes Sauvages des Lacs, Rivières et Tourbieres**. Saint-Augustin, Québec: Fleurbec, 1987. 399p. (Guide d'Identification Fleurbec). ISBN 292017410Xpa.
     65 species; color photos; natural history. In French. Aquatic plants of lakes, rivers, and bogs of Québec.

221. Lamoureux, Gisele, ed. **Plantes Sauvages des Villes et des Champs**. Saint-Cuthbert, Québec: Fleurbec, 1977. 273p. (Guide d'Identification Fleurbec). ISBN 2920174002.
     Not seen; see entry 222.

222. Lamoureux, Gisele, ed. **Plantes Sauvages des Villes, des Champs et en Bordure des Chemins, 2**. Saint-Augustin, Québec: Fleurbec, 1983. 208p. (Guide d'Identification Fleurbec). ISBN 292017407Xpa.
     62 species; color photos; maps. Includes wildflowers and grasses arranged by color of flowers. Descriptions include multiple common names in French and English, similar species, habitat and distribution, use for food, medicine, and forage. Covers the plants of Quebec. In French.

223. Leblanc, Danielle, et al. **Petite Flore Forestière du Québec**. 2e ed. rev. et augm. Québec: Publications du Québec, 1990. 249p. ISBN 2551122651.
     300 species; color photos. In French. A little forest flora of Québec.

224. Legasy, Karen L. **Field Guide to Forest Plants of Northeastern Ontario**. Timmins: Northeast Science and Technology, Ontario Ministry of Natural Resources, 1995. (NEST Field Guide, FG–004). ISBN 0777837439.
     300 species of lichens, ferns, liverworts, trees, shrubs, herbs, and grasses; keys; color photos and line drawings. Illustrated by Shayna LaBelle-Beadman.

225. Lonard, Robert I. **Guide to Grasses of the Lower Rio Grande Valley, Texas**. Edinburg, TX: University of Texas-Pan American Press, 1993. 240p. $18.95pa. ISBN 0938738089pa.

     183 species; keys; black–and–white illustrations; uses technical language. Illustrated by Norman A. Browne and Ardath L. Egle.

226. Magee, Dennis W. **Freshwater Wetlands: A Guide to Common Indicator Plants of the Northeast**. Amherst: University of Massachusetts Press, 1981. 245p. $17.95pa. ISBN 0870233165; 0870233173pa.

     175 species; keys; line drawings. Includes grasses, sedges, and rushes. Illustrated by Abigail Rorer.

227. Mason, Charles T., Jr., and Patricia B. Mason. **A Handbook of Mexican Roadside Flora**. Tucson: University of Arizona Press, 1987. 380p. ISBN 0816509972pa.

     200 species; line drawings.

228. McCoy, Doyle. **Roadside Wild Fruits of Oklahoma**. Norman: University of Oklahoma Press, 1980. 82p. $12.95pa. ISBN 08061212556; 0806116269pa.

     105 species; color photos.

229. Miller, Dorcas S. **Winter Weed Finder: A Guide to Dry Plants in Winter**. Berkeley, CA: Nature Study Guild, 1989. 61p. $3.00pa. ISBN 0912550171pa.

     64 species; key; line drawings; pocket size. Identifies wild, non-woody plants with pods or dried flowers in winter ranging over eastern North America. Illustrated by Ellen Amendolara.

230. Newmaster, Steven G., Allan Harris, and Linda J. Kershaw. **Wetland Plants of Ontario**. Edmonton, Alta.: Lone Pine, 1997. 240p. $19.95pa. ISBN 1551050595pa.

     475 species; color photos and line drawings; arranged by type of plant.

231. Powell, A. Michael. **Grasses of the Trans-Pecos and Adjacent Areas**. Austin: University of Texas Press, 1994. 377p. $29.95pa; $75.00. ISBN 0292765568pa; 0292765533.

     268 species; field manual; keys; line drawings; covers Texas, New Mexico, Oklahoma, Arizona, and northern Mexico. Illustrated by Patricia R. Manning.

232.   Ramey, V. **Aquatic Plant Identification Deck**. Gainesville, FL: Center for Aquatic Plants, University of Florida, 1995. 67 leaves. $10.00.

    67 species; color photos; 3x4 inches deck of laminated cards, held together with a metal post. Color photo on the front and identification information on the back.

233.   Redington, Charles B. **Plants in Wetlands**. Dubuque, IA: Kendall/ Hunt, 1994. 393p. (Redington Field Guides to Biological Interactions). $30.95pa. ISBN 0840389833pa.

    100 species; line drawings; emphasis on interactions with other plants and animals of the eastern U.S. Spiralbound.

234.   Richardson, Alfred. **Plants of the Rio Grande Delta**. Austin: University of Texas Press, 1995. 440p. (Treasures of Nature Series). $45.00; $24.95pa. ISBN 0292770685; 0292770707pa.

    Covers 823 species exclusive of grasses; keys; line drawings and color photos; field manual. Revised edition of the author's *Plants of Southernmost Texas*.

235.   Silberhorn, Gene M. **Common Plants of the Mid-Atlantic Coast: A Field Guide**. Baltimore: Johns Hopkins University Press, 1999. 306p. $45.00; $16.95pa. ISBN 0801860806; 0801860814pa.

    150 species; key; line drawings; covers area from Long Island, New York, to Cape Fear, North Carolina. Illustrated by Mary Warinner.

236.   Stalter, Richard. **Barrier Island Botany**. Dubuque, IA: Wm. C. Brown, 1992. 134p. $13.95pa. ISBN 0697204294pa.

    100 common species; black-and-white illustrations; natural history; arranged by community. Includes information on the use of native vegetation for coastal plantings. Covers the northeastern U.S. from Assateague Island, Virginia, to Cape Cod.

237.   Stalter, Richard. **Barrier Island Botany: The Southeastern United States**. Dubuque, IA: Wm. C. Brown, 1993. 164p. $13.95pa. ISBN 0697212793pa.

    100 common species; black-and-white illustrations; natural history; arranged by community. Includes information on the use of native vegetation for coastal plantings.

238.   Stensaas, Mark. **Canoe Country Flora: Plants and Trees of the North Woods and Boundary Waters**. Duluth, MN: Pfeifer-Hamilton, 1996. 209p. $14.95pa. ISBN 1570251215pa.

96 species of plants from lichens to trees; line drawings; numerous common names. Oversized, more of a natural history than a field guide. Companion to the author's *Canoe Country Wildlife*. Illustrated by Jeff Sonstegard.

239. Stubbendieck, James L., Stephan L. Hatch, and Charles H. Butterfield. **North American Range Plants**. 5th ed. Lincoln: University of Nebraska Press, 1997. 510p. $55.00; $29.95pa. ISBN 0803242603; 0803292430pa.

   20 species; line drawings; maps; oversized. Designed for the use of ecologists and land managers. Descriptions include inflorescence, vegetative, and growth characteristics; forage value; and habitat.

240. Threlkeld, Niki. **Flowering Plants of Devon Island, Canadian High Arctic**. Las Cruces, NM: Flora, 1991. 87p.

   44 species; color illustrations; pocket sized. Alternate title, *Flowering Plants of the High-Arctic*. Covers the Northwest Territories.

241. Tiner, Ralph W. **A Field Guide to Coastal Wetland Plants of the Northeastern United States**. Amherst: University of Massachusetts, 1987. 285p. $18.95pa. ISBN 0870235370; 0870235389pa.

   160 species; keys; line drawings; includes list of coastal wetland sites. Covers the coast from Maine to Maryland. Illustrated by Abigail Royer.

242. Tiner, Ralph W. **Field Guide to Coastal Wetland Plants of the Southeastern United States**. Amherst: University of Massachusetts Press, 1993. 328p. $18.95pa; $50.00. ISBN 0870238337pa; 0870238329.

   250 species described and keyed, another 200 mentioned; keys; line drawings. Covers Atlantic and Gulf Coasts from Virginia to Texas. Illustrated by Abigail Royer.

243. Tull, Delena, and George Miller. **A Field Guide to Wildflowers, Trees, and Shrubs of Texas**. Austin: Texas Monthly, 1991. 344p. (A Texas Monthly Field Guide). $21.95pa. ISBN 0877191972; 0877191956pa.

   650 species; keys; color photos and line drawings; maps. Descriptions include range and related species. Includes cacti.

244. Uva, Richard H., Joseph C. Neal, and Joseph M. DiTomaso. **Weeds of the Northeast**. Ithaca, NY: Comstock, 1997. 397p. $60.00; $29.95pa. ISBN 0801433916; 0801483344pa.

299 species; vegetative key and "shortcut identification tables"; color photos; includes fold-out grass identification table. Includes extensive descriptions (one species per page) with facing photos of weed parts; no information on control. Covers from Maine and southeastern Canada south to Virginia and west to Wisconsin.

245.   Van Bruggen, Theodore. **Wildflowers, Grasses and Other Plants of the Northern Plains and Black Hills**. 4th ed. Interior, SD: Badlands Natural History Association, 1992. 96p. $5.95.

312 representative species; color photos. Grasses and ferns are discussed first, followed by wildflowers, arranged by color.

246.   Weniger, Del. **Cacti of Texas and Neighboring States: A Field Guide**. Austin: University of Texas Press, 1984. 356p. ISBN 0292710852; 0292710631pa.

325 species; keys; color photos. Based on material from the author's *Cacti of the Southwest*.

247.   Wharton, Mary E., and Roger W. Barbour. **A Guide to the Wildflowers and Ferns of Kentucky**. Lexington: University Press of Kentucky, 1971. 344p. (Kentucky Nature Studies, No. 1). $27.95. ISBN 0813112346.

482 species illustrated by color photos, with another 180 species described. Illustrations and species descriptions in separate sections.

Western North America

248.   Archibald, J. H., G. D. Klappstein, and I. G. W. Corns. **Field Guide to Ecosites of Southwestern Alberta**. Edmonton, Alta.: Northern Forestry Centre, 1996. (Special Report of the Northern Forestry Centre of Canada No. 5). ISBN 066016440X (Binder).

112 species; color photos and black-and-white illustrations; no descriptions. The primary objective of this book is to classify and describe about 100 ecosystems of the region, plus 17 soil types. The authors also include photos of common plant species.

249.   Atkinson, Scott, and Fred Sharpe. **Wild Plants of the San Juan Islands**. 2nd ed. Seattle, WA: Mountaineers/San Juan Preservation Trust, 1993. 191p. $12.95pa. ISBN 0898863562pa.

190 species of flowers, trees, shrubs, grasses; line drawings; checklist of all 800 species found in the islands. Arranged by habitat.

250. Batson, Wade T. **Genera of the Western Plants**. Columbia, SC: Printed by State Printing, 1982. 207p.

Comprehensive; key only; line drawings.

251. Becking, Rudolf Willem. **Pocket Flora of the Redwood Forest**. Covelo, CA: Island, 1982. 237p. $16.95pa. ISBN 0933280025pa.

212 species; keys; line drawings, some color photos; arranged by family.

252. Beckingham, John D., and J. H. Archibald. **Field Guide to Ecosites of Northern Alberta**. Edmonton, Alta.: Northern Forestry Centre (Canada), 1996. 1 v. (various paging). (Special Report of the Northern Forestry Centre, No. 5).

87 species; color photos and black-and-white illustrations; no descriptions. The primary objective of this book is to classify and describe about 100 ecosystems of the region, plus 17 soil types. The authors also include photos of common plant species.

253. Beckingham, John D., I. G. W. Corns, and J. H. Archibald. **Field Guide to Ecosites of West-Central Alberta**. Edmonton, Alta.: Northern Forestry Centre, 1996. (Special Report of the Northern Forestry Centre, No. 9). ISBN 0660164418 (bound); 0660164426 (binder).

106 species; color photos and black-and-white illustrations; no descriptions. The primary objective of this book is to classify and describe about 100 ecosystems of the region, plus 17 soil types. The authors also include photos of common plant species.

254. Beckingham, John D., D. G. Nielsen, and V. A. Futoransky. **Field Guide to Ecosites of the Mid-Boreal Ecoregions of Saskatchewan**. Edmonton, Alta.: Canadian Forest Service, 1996. 1 v. (various paging). (Special Report of the Northern Forestry Centre, No. 6). ISBN 066016387X; 0660163888.

103 species; color photos and black-and-white illustrations; no descriptions. The primary objective of this book is to classify and describe about 100 ecosystems of the region, plus 17 soil types. The authors also include photos of common plant species.

255. Benson, Lyman David. **The Native Cacti of California**. Stanford, CA: Stanford University Press, 1969. 243p. $14.95pa. ISBN 0804715262pa.

Field manual; keys; black-and-white photos and illustrations, some color; maps; extensive natural history.

256.   Burt, Page. **Barrenland Beauties: Showy Plants of the Arctic Coast**. Yellowknife, N.W.T.: Outcrop, 1991. 246p. ISBN 0919315259.

   100 species; key by color and shape; color photos; also in Inuinaktun (local dialect of Inuit).

257.   Collins, Barbara J. **Key to Coastal and Chaparral Flowering Plants of Southern California**. 2nd ed. Dubuque, IA: Kendall/Hunt, 1996. 336p. $31.95pa. ISBN 0840343299pa.

   Key; line drawings.

258.   Cooke, Sarah Spear, ed. **A Field Guide to the Common Wetland Plants of Western Washington and Northwestern Oregon**. Seattle, WA: Seattle Audubon Society, 1997. 417p. (Trailside Series). ISBN 0914516116pa.

   300 species of ferns, grasses, sedges, trees, shrubs, and flowers; keys; black-and-white illustrations and color photos; maps. Descriptions include habitat, range, and similar species. Descriptions of genera include forage value and ethnobotanical use.

259.   Crampton, Beecher. **Grasses in California**. Berkeley, CA: University of California Press, 1974. 178p. (California Natural History Guide, No. 33). $12.95pa. ISBN 0520027396; 0520025075pa.

   162 species; keys; line drawings.

260.   Dawson, Elmer Yale, and Michael S. Foster. **Seashore Plants of California**. Berkeley, CA: University of California Press, 1982. 226p. (California Natural History Guides, No. 47). $10.95pa. ISBN 0520041380; 0520041399pa.

   240 species; keys; line drawings and color photos. Combination and revision of *Seashore Plants of Northern California* and *Seashore Plants of Southern California*, both by E. Yale Dawson, 1966. Illustrated by Bruce Stewart.

261.   Earle, W. Hubert. **Cacti of the Southwest: Arizona, Western New Mexico, Southern Colorado, Southern Utah, Southern Nevada, Eastern California**. Rev. ed. Phoenix, AZ: Rancho Arroyo, 1980. 210p. ISBN 0935810307; 0935810315pa.

   152 species; key; color and black-and-white photos.

262.   Engard, Rodney G. **The Flowering Southwest: Wildflowers, Cacti, and Succulents in Arizona, California, Colorado, Nevada, New Mexico, Texas, and Utah**. Tucson, AZ: Great Impressions, 1989. 120p. $16.95. ISBN 0925263001.

   50 species; color illustrations; natural history. More of an art book than a field guide. Illustrated by Erni Cabat.

263. Epple, Anne Orth, and Lewis E. Epple. **A Field Guide to the Plants of Arizona**. Mesa, AZ: Helena, MT: LewAnn; Distributed by Falcon, 1995. 347p. $24.95pa. ISBN 1560443146pa.

    850 species; color photos are separate from the text; flowers are arranged by color, other plants are arranged by form.

264. Faber, Phyllis M., and Robert F. Holland. **Common Riparian Plants of California: A Field Guide for the Layman**. Mill Valley, CA: Pickleweed, 1988. 140p. $18.00pa. ISBN 0960789014pa.

    120 species; life-sized photocopies of plants.

265. Faber, Phyllis M., and Robert F. Holland. **Common Wetland Plants of Coastal California: A Field Guide for the Layman**. Mill Valley, CA: Pickleweed, 1982. 140p. $15.00pa. ISBN 0960789006pa.

    95 species; life-sized photocopies of plants.

266. Fagan, Damian. **Canyon Country Wildflowers: A Field Guide to 190 Wildflowers, Shrubs and Trees**. Helena, MT: Falcon, 1998. 192p. $16.95pa. ISBN 1560445602pa.

    Not seen. 190 species; color photos. Covers the Four Corners area.

267. Fischer, Pierre C. **70 Common Cacti of the Southwest**. Tucson, AZ: Southwest Parks and Monuments Association, 1989. 80p. $6.95pa. ISBN 0911408827pa.

    70 species; color photos.

268. Guard, B. Jennifer, and John Christy. **Wetland Plants of Oregon and Washington**. Redmond, WA: Lone Pine, 1995. 239p. $19.95pa. ISBN 1551050609pa.

    350 species; 5 keys; color photos and line drawings; natural history; arranged by habitat. Descriptions include leaves, flowers, fruit, similar species, and habitat.

269. Hall, Judy Kathryn. **Native Plants of Southeast Alaska**. Juneau, AK: Windy Ridge, 1995. 284p. $24.95. ISBN 0965872602.

    500 species of ferns, trees, shrubs, grasses, sedges, and wildflowers; keys; line drawings; includes information on use. Spiral-bound. Illustrated by Claudia Kelsey.

270. Hallworth, Beryl, and C. C. Chinnappa. **Plants of Kananaskis Country in the Rocky Mountains of Alberta**. Edmonton, Alta.: University of Alberta Press, 1997. 368p. ISBN 0888642970pa.

    423 species; keys; black-and-white illustrations and color photos; checklist. Illustrated by Sharon Orser.

271. Hughes, Jeffrey W., and Will H. Blackwell. **Wildflowers (and Other Plant Life) of Southeast Alaska: An Identification Guide.** Dubuque, IA: Kendall/Hunt, 1987. 123p. ISBN 0840344511pa.

     75 species; line drawings; spiralbound. Illustrated by Elizabeth Mills and Valerie Sloan.

272. Johnson, Derek, Trevor Goward, and Dale H. Vitt. **Plants of the Western Boreal Forest and Aspen Parkland.** Redmond, WA: Lone Pine, 1995. 392p. $19.95pa. ISBN 1551050587pa.

     1,350 plants; keys, including color key to flowers; color photos and line drawings; includes information on use.

273. Johnson, Paul R. **Cacti, Shrubs, and Trees of Anza-Borrego: An Amateur's Key for Identifying Desert Plants.** Borrego Springs, CA: Anza-Borrego Desert Natural History Association, 1990. 28p. $3.50pa. ISBN 0910805067pa.

     90 species; key; line drawings. Pamphlet.

274. Jones, Alice Goen. **Flowers and Trees of the Trinity Alps: A Photographic Guide.** Weaverville, CA: Trinity County Historical Society, 1986. 195p. ISBN 0960705422pa.

     332 species; color photos in separate plates; arranged by color. Covers the Trinity County area of northern California.

275. Keator, Glenn. **Plants of the East Bay Parks.** Niwot, CO: Roberts Rinehart, 1994. 332p. $19.95pa. ISBN 1879373424pa.

     1,500 species; keys; line drawings; sites. Covers Alameda and Contra Costa counties in California. Illustrated by Peg Steunenberger and Susan Bazell.

276. Kershaw, Linda J. **Plants of the Rocky Mountains.** Edmonton, Alta.: Lone Pine, 1997. $19.95pa. ISBN 1551050889pa.

     1,362 species of trees, shrubs, wildflowers, grasses, ferns, mosses, and lichens; color photos and black-and-white illustrations. Descriptions include habitat and distribution, general notes, use, and edibility.

277. Knute, Adrienne. **Plants of the East Mojave.** Cima, CA: Wide Horizons, 1991. 207p. $12.95pa. ISBN 0938109081.

     175 species; black-and-white photos.

278. Komarek, Susan. **Flora of the San Juans: A Field Guide to the Mountain Plants of Southwestern Colorado.** Durango, CO: Kivaki, 1994. 244p. $18.95pa. ISBN 1882308069pa.

     700 species; key throughout; line drawings and color illustrations.

279. Kozloff, Eugene N., and Linda H. Beidleman. **Plants of the San Francisco Bay Region: Mendocino to Monterey**. Pacific Grove, CA: Sagen, 1994. 332p. $29.95pa. ISBN 0964375605pa.

     2,000 species; key throughout; color photos and line drawings. Includes grasses and ferns.

280. Leake, Dorothy Van Dyke, John Benjamin Leake, and Marcelotte Leake Roeder. **Desert and Mountain Plants of the Southwest**. Norman: University of Oklahoma Press, 1993. 239p. $19.95pa. ISBN 080612489Xpa.

     253 species; line drawings and color photos. Includes mosses, ferns, trees, and flowers; flowers are arranged by color. Illustrated by Dorothy Van Dyke Leake.

281. Lyons, C. P. **Trees, Shrubs and Flowers to Know in Washington and British Columbia**. Edmonton, Alta.: Lone Pine, 1995. $15.95pa. ISBN 1551050625pa.

     600 species; line drawings and color photos; maps. Wildflowers arranged by color. Also published as *Trees, Shrubs and Flowers to Know in British Columbia and Washington*. Expanded edition of the author's *Trees, Shrubs and Flowers to Know in British Columbia*.

282. Lyons, Kathleen M., and Mary Beth Cuneo-Lazaneo. **Plants of the Coast Redwood Region**. Los Altos, CA: Looking, 1988. 197p. $15.00pa. ISBN 0962696102pa.

     185 species; color photos; arranged by group and flower color.

283. MacKinnon, Andy, Jim Pojar, and Ray Coupe, eds. **Plants of Northern British Columbia**. Edmonton, Alta.: Lone Pine, 1992. 344p. $19.95pa. ISBN 1551050153pa.

     640 species of trees, shrubs, wildflowers, grasses, ferns, mosses, and lichens; keys; color photos and line drawings.

284. Mielke, Judy. **Native Plants for Southwestern Landscapes**. Austin: University of Texas Press, 1993. 384p. $39.95; $24.95pa. ISBN 0292755538; 0292751478pa.

     280 species of trees, shrubs, vines, grasses, groundcovers, perennials, cacti, and succulents; color photos; arranged by family. Not really a field guide, but rather an encyclopedia of plants suitable for landscape use. Field guide type entries include description, distribution, propagation, and landscape use. A good source of information on desert plants and small enough to carry in the field despite the horticultural orientation.

285. Nelson, Ruth Ashton. **Handbook of Rocky Mountain Plants**. 4th ed. Revised by Roger L. Williams. Niwot, CO: Roberts Rinehart, 1992. 444p. $19.95pa. ISBN 0911797963pa.

Comprehensive; keys; line drawings. Illustrated by Dorothy Van Dyke Leake.

286. Nelson, Ruth Ashton. **Plants of Rocky Mountain National Park**. 5th ed. Estes Park, CO: Rocky Mountain Nature Association, 1982. 168p. ISBN 0870811266.

750 species; keys; black-and-white photos.

287. Nelson, Ruth Ashton. **Plants of Zion National Park: Wildflowers, Trees, Shrubs, and Ferns**. Springdale, UT: Zion Natural History Association, 1976. 333p. ISBN 0915630001; 091563001Xpa.

700 species; keys; color photos and black-and-white illustrations; arranged by seasons of blooming and by color. Illustrated by Tom Blaue.

288. Parish, Roberta, R. Coupe, Dennis Lloyd, and Joe Antos. **Plants of Southern Interior British Columbia**. Vancouver, B.C.: Lone Pine, 1996. 463p. $19.95pa. ISBN 1551050579pa.

700 species of ferns, mosses, lichens, trees, shrubs, grasses, and wildflowers; keys; line drawings and color photos. Covers area from the Rocky Mountains west to the coastal mountains of British Columbia, Washington, and Idaho.

289. Pesman, M. Walter. **Meet the Natives: A Beginner's Field Guide to Rocky Mountain Wildflowers, Trees, and Shrubs**. 9th ed. Boulder, CO: Pruett, 1992. 248p. $12.95pa. ISBN 1879373319pa.

400 species; line drawings; arranged by altitude, flower color, and family.

290. Pojar, Jim, and Andy MacKinnon. **Plants of the Pacific Northwest Coast: Washington, Oregon, British Columbia and Alaska**. Redmond, WA: Lone Pine, 1994. 527p. $19.95pa. ISBN 1551050404pa.

794 species; keys; color photos and line drawings; maps; description, natural history, notes. Includes flowering plants, grasses, sedges, rushes, ferns, mosses, liverworts, and lichens. Also published as *Plants of Coastal British Columbia*.

291. Roberts, Norman C. **Baja California Plant Field Guide**. La Jolla, CA: Natural History Publications, 1989. 309p. $24.95pa. ISBN 0960314415pa.

550 plants are described, 270 of which are illustrated; key to families; color photos.

292. Shaw, Richard J. **Field Guide to the Vascular Plants of Grand Teton National Park and Teton County, Wyoming**. Logan: Utah State University Press, 1976. 301p. ISBN 087421081X.

900 species; field guide/manual; key; black-and-white illustrations.

293. Shaw, Richard J. **Plants of Yellowstone and Grand Teton National Parks**. Rev. ed. Salt Lake City, UT: Wheelwright, 1981. 159p. ISBN 0937512028.

213 species of ferns, trees, and flowers; separate keys for coniferous and deciduous trees; color photos; flowers arranged by color.

294. Shaw, Richard J., and Danny On. **Plants of Waterton-Glacier National Parks and the Northern Rockies**. Missoula, MT: Mountain, 1979. 160p. $12.00pa. ISBN 0878421149pa.

200 species; color photos; arranged by color.

295. Sohmer, S. H. **Plants and Flowers of Hawaii**. Honolulu: University of Hawaii Press, 1987. 160p. $17.95. ISBN 0824810961.

151 species; color photos; arranged by habitat; extensive background and natural history.

296. Taylor, Ronald J. **Northwest Weeds: The Ugly and Beautiful Villains of Fields, Gardens, and Roadsides**. Missoula, MT: Mountain, 1990. 177p. $14.00pa. ISBN 0878422498pa; 0878422609.

240 species; color photos; natural history; no weed control information is provided.

297. Taylor, Ronald J., and George W. Douglas. **Mountain Plants of the Pacific Northwest: A Field Guide to Washington, Western British Columbia, and Southeastern Alaska**. Missoula, MT: Mountain, 1995. 437p. $20.00pa. ISBN 0878423141pa.

450 species of ferns, trees, forbs, grasses; keys; color photos.

298. Tierney, Gail D. **Roadside Plants of Northern New Mexico**. Santa Fe, NM: Lightning Tree, 1983. 175p. ISBN 0890160678; 0890160619pa.

122 species; line drawings, color illustrations; by habitat and seasons. Illustrated by Phyllis Hughes.

299. Weber, William A. **Rocky Mountain Flora: A Field Guide for the Identification of the Ferns, Conifers, and Flowering Plants of the Southern Rocky Mountains from Pikes Peak to Rocky Mountain National Park and from the Plains to the Continental Divide**. 5th ed., rev. Boulder, CO: Colorado Associated University Press, 1976. 479p. $20.00pa. ISBN 0685643263; 0870810685pa.

      750 species; keys; black-and-white illustrations; field guide/manual. Originally published as *Handbook of Plants of the Colorado Front Range*.

300. Weber, William A., and Ronald C. Wittman. **Colorado Flora: Eastern Slope**. Rev. ed. Niwot, CO: University Press of Colorado, 1996. 524p. $32.50pa. ISBN 0870813870pa.

      Comprehensive; keys; line drawings and some color photos; alphabetical by family name.

301. Weber, William A., and Ronald C. Wittman. **Colorado Flora: Western Slope**. Rev. ed. Boulder, CO: Colorado Associated University Press, 1996. 496p. $32.50pa. ISBN 0870813889pa.

      Comprehensive; keys; line drawings and color illustrations; alphabetical by family.

302. Weeden, Norman. **A Sierra Nevada Flora**. 4th ed. Berkeley, CA: Wilderness, 1996. 259p. $15.95pa. ISBN 0899972047pa.

      700 species; keys; black-and-white illustrations; technical language (includes glossary); edibility information is provided for most species covered.

303. Wiedemann, Alfred M., La Rea J. Dennis, and Frank H. Smith. **Plants of the Oregon Coastal Dunes**. Corvallis, OR: O.S.U. Book Stores, 1974. 117p. $7.50pa. ISBN 0882461176pa.

      90 species; keys; black-and-white photos; extensive background information. Spiralbound.

Edible and Useful Plants

Unlike most field guides, the guides covered in this chapter have a practical purpose: to assist users in identifying plants they can use or should avoid. This includes edible, medicinal, and poisonous plants, plus a variety of plants put to other uses such as for dyeing fabric. Many of the edible plant guides include a few recipes or tips on using the plants. A few books covering cultivated plants such as garden herbs or exotic vegetables are included, but most are for plants growing in the wild. There are a number of books that list plants and discuss their uses that are not included here, since their primary purpose is not to identify the plants, but to document their use.

Caution should be used whenever dealing with edible or medicinal plants. Misidentifying or misusing plants can be dangerous. Wild stocks of some popular species such as ginseng or echinacea (purple coneflower) have been decimated by indiscriminate collecting, so the prudent collector should not collect rare or threatened plants.

Many general plant guides found in Chapter 2, "Flora and Fauna," and Chapter 3, "Plants," also include information about plant uses. In addition, poisonous mushrooms are identified in almost all mushroom guides (see Chapter 6).

NORTH AMERICA

304. Andrews, Jean. **The Pepper Lady's Pocket Pepper Primer**. Austin: University of Texas Press, 1997. 184p. $17.95pa. ISBN 0292704828; 0292704836pa.

42 species or varieties; color photos; descriptions include taste, use, sources, substitutions, other names, and general remarks. Also includes general information on using and growing peppers, pepper nomenclature, and other topics.

❑   Angell, Madeline. **A Field Guide to Berries and Berrylike Fruits**. See entry 177.

305.   Angier, Bradford. **Field Guide to Edible Wild Plants**. Harrisburg, PA: Stackpole, 1974. 256p. $16.95pa. ISBN 0811706168; 0811720187pa.

120 species; color illustrations, description, range, edibility; emphasis on use, not identification.

306.   Angier, Bradford. **Field Guide to Medicinal Wild Plants**. Harrisburg, PA: Stackpole, 1978. 320p. $18.95pa. ISBN 0811720764pa.

300 species; color illustrations; use; emphasis on identification. Illustrated by Arthur J. Anderson.

307.   Bliss, Anne. **North American Dye Plants**. New York: Scribner, 1980. 288p. ISBN 0684163934.

Not exactly a field guide, but contains drawings, descriptions, and uses of 75 species. Also includes colors obtained by using different mordants in the dyeing process. Revised and enlarged edition of *Rocky Mountain Dye Plants*. Illustrated by Robert Bliss.

308.   Burger, Sandra and A. P. Knight. **Horse Owner's Field Guide to Toxic Plants**. Ossining, NY: Breakthrough, 1996. 230p. $22.50pa. ISBN 0914327623pa.

100 toxic species are described and illustrated, another 11 suspected species described; color photos; maps; arranged by type of plant. Descriptions include multiple common names, similar species, distribution, signs of poisoning, and what to do if poisoning is suspected.

309.   Dawson, Ron. **Nature Bound: A Pocket Field Guide**. Boise, ID: OMNIgraphics, 1985. 335p. $12.00pa. ISBN 0960977678pa.

80 species of edible and 34 species of poisonous plants; color photos; includes preparation information for edible plants and symptoms and treatment for poisonous ones; pocket-sized. While the title does not indicate it, this is actually a survival guide. The introductory material provides brief information on starting a fire, creating a shelter, how to tie knots, how to set snares and skin mammals, first aid, and so on. The poisonous and edible plant section covers about two-thirds of the pages, with a two-page spread for each plant.

310. Elias, Thomas S., and Peter A. Dykeman. **Edible Wild Plants: A North American Field Guide**. New York: Sterling, 1990. 286p. $16.95pa. ISBN 0806974885pa.

200 common species; seasonal key; color photos; arranged by seasons and types. Reprint. Originally published in 1982.

311. Elliott, Douglas B. **Wild Roots: A Forager's Guide to the Edible and Medicinal Roots, Tubers, Corms, and Rhizomes of North America**. Rochester, VT: Healing Arts, 1995. 128p. $14.95pa. ISBN 0892815388pa.

70 species; line drawings of roots alone (not full plants); includes natural history and use. Slightly oversized for a field guide, but useful for confirming the identification of a root once the plant is identified using a standard field guide. Originally published as *Roots*. Illustrated by the author.

312. Forey, Pamela. **An Instant Guide to Medicinal Plants: The Medicinal Plants of North America and their Uses Described and Illustrated in Full Color**. New York: Crescent, 1991. 123p. (Instant Guide Series). $4.99. ISBN 0517691132.

110 species; color illustrations; checklist.

313. Forey, Pamela, and Cecilia Fitzsimons. **An Instant Guide to Edible Plants: The Most Familiar Edible Wild Plants of North America**. New York: Bonanza, 1989. 123p. (Instant Guide Series). $4.99. ISBN 0517662175.

100 species; color illustrations; maps; checklist. Arranged by edible part of the plant.

❑ Foster, Steven, and Roger A. Caras. **A Field Guide to Venomous Animals and Poisonous Plants: North America North of Mexico**. See entry 108.

314. Furlong, Marjorie, and Virginia Pill. **Wild Edible Fruits and Berries**. Healdsburg, CA: Naturegraph, 1974. 64p. $8.95pa. ISBN 0879610336; 0879610328pa.

42 species; color photos; 20 recipes.

315. James, Wilma Roberts. **Know Your Poisonous Plants: Poisonous Plants Found in Field and Garden**. Healdsburg, CA: Naturegraph, 1973. 99p. $7.95pa. ISBN 0879610131; 0879610123pa.

66 species covered in some detail, plus brief discussions of another 75 species; line drawings; includes description and signs of poisoning. In alphabetical order by common name. Illustrated by Arla Lippsmeyer.

316.   Krochmal, Arnold, and Connie Krochmal. **A Field Guide to Medicinal Plants**. Updated Times Books paperback ed. New York: Times, 1984. 274p. $15.00pa. ISBN 0812963369pa.

   270 species; arranged by genus; black and white drawings and photos; common names, descriptions, use, and range.

317.   Muenscher, Walter Conrad Leopold. **Poisonous Plants of the United States**. Rev. ed. New York: Collier, 1975. 277p.

   400 species; black and white illustrations; arranged by family with description, range, symptoms, and toxicity. Illustrated by Helen Hill Craig.

318.   Schultes, Richard Evans. **Hallucinogenic Plants**. New York: Golden, 1976. 160p. ISBN 0307243621pa; 030764362X.

   75 species; color illustrations; includes use by traditional cultures, and chemistry. Illustrated by Elmer W. Smith.

319.   Springmeyer, Fritz. **Willow Bark and Rosehips: An Introduction to Common Edible and Useful Wild Plants of North America**. Helena, MT: Falcon, 1996. 80p. $9.95. ISBN 1560444126pa.

   30 species; color illustrations. A colorful guide listing habitat, description, uses, a few recipes, plus general notes on collecting and using plants. Illustrated by Michele Montez.

320.   Szczawinski, Adam F., and Nancy J. Turner. **Edible Garden Weeds of Canada**. Ottawa: National Museum of Natural Sciences, 1978. 184p. (Edible Wild Plants of Canada, No. 1). ISBN 0660000261 (English ed.); 066000027X (French ed.).

   35 species; color photos and black and white illustrations; multiple common names; descriptions include use, location, and three to five recipes per plant.

321.   Tatum, Billy Joe. **Billy Joe Tatum's Wild Foods Cookbook and Field Guide**. New York: Workman, 1976. 268p. $7.95pa. ISBN 0911104763; 0911104771pa.

   70 species; line drawings; separate cookbook section featuring 350 recipes. Illustrated by Jim Blackfeather Rose.

322.   Turner, Nancy J., and Adam F. Szczawinski. **Common Poisonous Plants and Mushrooms of North America**. Portland, OR: Timber, 1991. 311p. $55.00; $24.95pa. ISBN 0881921793; 0880923125pa.

   16 species of mushrooms, 135 species of plants; color photos; background, identification, toxicity and treatment for each species. Mushrooms and plants are in separate sections; plants are arranged by habitat.

Eastern North America

323. Foster, Steven, and James A. Duke. **A Field Guide to Medicinal Plants: Eastern and Central North America**. Boston: Houghton Mifflin, 1998. 366p. (Peterson Field Guide Series, No. 40). $18.00pa. ISBN 0395920663pa.

    500 species; line drawings, color illustrations and color photos; uses. Illustrated by Roger Tory Peterson, Jim Blackfeather Rose, and Lee Allen Peterson.

324. Kavasch, E. Barrie. **Guide to Northeastern Wild Edibles**. Blaine, WA: Hancock House, 1981. 64p. $6.95pa. ISBN 0888390904pa.

    95 species; color photos and line drawings; arranged by season.

325. Krumm, Bob. **The Great Lakes Berry Book: A Complete Guide to Finding, Harvesting, and Preparing Wild Berries and Fruits in the Great Lakes**. Helena, MT: Falcon, 1996. 143p. (Falcon Field Guide). $11.95pa. ISBN 156044886pa.

    22 species; color photos; 115 recipes. This is a cookbook with descriptions and natural history about the berries included. Each berry is introduced with a narrative covering habitat, description, season, and the author's favorite berrying trips.

326. Krumm, Bob. **The New England Berry Book: Field Guide and Cookbook**. Camden, ME: Yankee, 1990. 110p. $11.95pa. ISBN 0899092136pa.

    12 species; color photos; 100 recipes. Similar in format to *The Great Lakes Berry Book* (see entry 325).

327. Lamoureux, Gisele, ed. **Plantes Sauvages Comestibles**. Saint-Cuthbert, Québec: Fleurbec, 1981. 167p. ISBN 2920174037pa.

    16 poisonous plants and 28 edible species; color photos. Descriptions include common names in English and French, general notes, cultivation, uses for food and medicine, similar species, and habitat. Also includes color photos and brief descriptions of an additional 44 species of edible plants covered in the companion volumes, *Plantes Sauvages Printanières* (see entry 577) and *Plantes Sauvages des Villes et des Champs* (see entry 221), including references to page numbers in those books. In French. Covers the edible plants of Quebec. *Plantes Sauvages au Menu* is a companion cookbook, containing recipes, cooking methods, and nutritive values.

328.   Miller, Dorcas S. **Berry Finder: A Guide to Native Plants with Fleshy Fruits for Eastern North America**. Berkeley, CA: Nature Study Guild, 1986. $3.00pa. ISBN 0912550228pa.

140 species; key throughout; line drawings; pocket size.

329.   Naegele, Thomas A. **Edible and Medicinal Plants of the Great Lakes Region**. Rev. ed. Davisburg, MI: Wilderness Adventure, 1996. 423p. $18.95pa. ISBN 0923568379pa.

150 species; black and white illustrations; includes descriptions and use. The author is listed as DO (Doctor of Osteopathy).

330.   Peterson, Lee Allen. **A Field Guide to Edible Wild Plants of Eastern and Central North America**. Boston: Houghton Mifflin, 1978. 330p. (Peterson Field Guide Series, No. 23). $26.00; $18.00pa. ISBN 0395204453; 039531870Xpa.

370 edible species and 37 poisonous look-alikes; line drawings and color photos; descriptions include where found, use, warnings, and general notes. Includes lists of edible plants found in 14 habitats and edible plants arranged by type of use. Has basic instructions for preparing foods, but no recipes. Illustrated by Lee Allen Peterson and Roger Tory Peterson.

331.   Stephens, H. A. **Poisonous Plants of the Central United States**. Lawrence, KS: Regents Press of Kansas, 1980. 165p. $15.95pa. ISBN 070060202X; 0700602046pa.

150 species; black and white photos; description and toxicity.

332.   TerBeest, Char. **Gifts from the Earth: A Basketmaker's Field Guide to Midwest Botanicals**. Baraboo, WI: Wild Willow Press; Cross Plains, WI: Ampersand, 1988. 47p. ISBN 0961479515pa.

29 species or groups of plants useful to basketmakers; black and white illustrations. The plants covered include both cultivated and wild species, and the text includes information on identifying, locating, and using the plants for basketmaking. Illustrated by Rhonda Ness.

Western North America

333.   **Alaska Wild Berry Guide and Cookbook**. Anchorage, AK: Alaska Northwest, 1982. 201p. $14.95pa. ISBN 0882402293pa.

50 species; color photos and line drawings; descriptions are in the front of the book, followed by 277 recipes.

334. Clarke, Charlotte Bringle. **Edible and Useful Plants of California**. Berkeley, CA: University of California Press, 1977. 280p. (California Natural History Guides, No. 41). $14.95pa. ISBN 0520032616; 0520032675pa.

    Line drawings; recipes, description, range; emphasis on use, not identification.

335. Fuller, Thomas C., and Elizabeth McClintock. **Poisonous Plants of California**. Berkeley, CA: University of California Press, 1986. 433p. (California Natural History Guides, No. 53). $13.95pa. ISBN 0520055683; 0520055691pa.

    1,240 species, from algae to flowering plants; color photos and line drawings; includes symptoms of poisoning.

336. Harrington, Harold David. **Edible Native Plants of the Rocky Mountains**. Albuquerque: University of New Mexico Press, 1972. 392p. $16.95pa. ISBN 0826303439pa.

    160 species; arranged by use; black and white illustrations; description, use. Emphasis is on use, not identification. Illustrated by Y. Matsumura.

337. Harrington, Harold David. **Western Edible Wild Plants**. Albuquerque, NM: University of New Mexico Press, 1967. 156p. ISBN 0826302181.

    125 species; black and white drawings; description, use; emphasis on use, not identification. Illustrated by Y. Matsumura.

338. Keator, Glenn. **Pacific Coast Berry Finder: A Pocket Manual for Identifying Native Plants with Fleshy Fruits**. Berkeley, CA: Nature Study Guild, 1978. 59p. $3.00pa. ISBN 0912550023pa.

    118 species; keys; black and white illustrations; maps. Pocket sized.

339. Kirk, Donald R. **Wild Edible Plants of Western North America**. Rev. ed. Happy Camp, CA: Naturegraph, 1975. 307p. $6.95pa. ISBN 0911010858; 091101084Xpa.

    2,000 species; black and white illustrations; arranged by location; brief description, use, range. Originally published as *Wild Edible Plants of the Western United States*. Illustrated by Janice Kirk.

340. Krumm, Bob. **The Rocky Mountain Berry Book**. Helena, MT: Falcon, 1991. 162p. $9.95pa. ISBN 156044046pa.

    15 species; color photos; 148 recipes. Similar in format to *The Great Lakes Berry Book*, entry 325.

341. Pratt, Verna E. **Alaska's Wild Berries and Berry-Like Fruit**. Anchorage, AK: Alaskakrafts, 1995. 127p. $9.95pa. ISBN 0962319244pa.

     50 species; color photos; arranged by edible species, followed by inedible species and by color within each group. Pocket sized.

342. Schofield, Janice J. **Alaska's Wild Plants: A Guide to Alaska's Edible Harvest**. Anchorage: Alaska Northwest, 1993. 95p. $12.95pa. ISBN 0882404334pa.

     70 species; color photos; includes edibility, medicinal use, and general notes; pocket sized.

343. Schofield, Janice J. **Discovering Wild Plants: Alaska, Western Canada, the Northwest**. Anchorage: Alaska Northwest, 1989. 354p. $32.95pa. ISBN 0882403559pa.

     147 species, from fungi to flowering plants; color photos and line drawings; arranged by habitat; includes information on use and some recipes. Oversized. Illustrated by Richard W. Tyler.

344. Sweet, Muriel. **Common Edible and Useful Plants of the West**. Healdsburg, CA: Naturegraph, 1976. 64p. $5.95pa. ISBN 0879610476; 0879610468pa.

     116 species; black and white illustrations; includes poisonous plants; emphasis on identification.

345. Tilford, Gregory L. **Edible and Medicinal Plants of the West**. Missoula, MT: Mountain, 1997. 240p. $21.00pa. ISBN 0878423591pa.

     250 species; color photos; descriptions include field identification, habitat and range, blooming time, edibility, medicinal uses, and toxicity. Covers Great Lakes and Great Plains area west to the Pacific Ocean.

346. Underhill, J. E. **Wild Berries of the Pacific Northwest**. Seattle, WA: Superior, 1974. 128p. $8.95pa. ISBN 0919654061pa.

     68 species; key; color photos and line drawings; includes poisonous and medicinal berries; recipes for jams and wine.

347. Viereck, Eleanor. **Alaska's Wilderness Medicines: Healthful Plants of the Far North**. Edmonds, WA: Alaska Northwest, 1987. 107p. $9.95pa. ISBN 0882403222pa.

     50 species; line drawings; includes description, use, constituents. Illustrated by Dominique Collet.

348. Wiltens, James S. **Thistle Greens and Mistletoe: Edible and Poisonous Plants of Northern California**. Berkeley, CA: Wilderness, 1988. 160p. $9.95pa. ISBN 0899970907pa.

50 species; black and white illustrations; multiple common names. Descriptions include habitat, uses, and general notes. Reprint was originally published as *Plants Your Mother Never Told You About*.

Mushrooms

Mushroom field guides are in many ways more similar to edible plant field guides than they are to standard plant guides. Almost all mushroom field guides include information about the edibility of their species, and many even include recipes. It seems that, unlike other field guide subjects where mere curiosity is assumed, mushroom experts assume that the only reason a person would be interested in learning about mushrooms is to dine on them. These semi-culinary guides may also go to some length to assure readers that eating mushrooms is not automatically fatal, despite their frequent appearance in murder mysteries. On the other hand, most guides stress the fact that readers should be very, very sure of their identification before dining.

Aside from this culinary emphasis, mushroom field guides are similar to other field guides. They are perhaps more likely to have photographs rather than illustrations, but both forms are found. There are relatively few regional guides; most mushroom guides cover a large area. One unique piece of information found in many mushroom guides is the spore print. The color of a mushroom's spores may be diagnostic, so many guides recommend checking the spore print before deciding on an identification.

Some mushrooms may also be included in the field guides listed in Chapter 2, "Flora and Fauna," and Chapter 3, "Plants."

NORTH AMERICA

349. Arora, David. **Mushrooms Demystified: A Comprehensive Guide to the Fleshy Fungi.** 2nd ed. Berkeley, CA: Ten Speed, 1986. 959p. $39.95; $35.00pa. ISBN 0898151708; 0898151716 (deluxe); 0898151694pa.

     2,000 species; keys; mostly black-and-white photos; natural history.

350. Bessette, Alan, Orson K. Miller, Jr., Arleen R. Bessette, and Hope H. Miller. **Mushrooms of North America in Color: A Field Guide Companion to Seldom-Illustrated Fungi.** Syracuse, NY: Syracuse University Press, 1995. 172p. $39.95; $17.95pa. ISBN 0815626665; 0815603231pa.

     75 species rarely included in mushroom field guides of common species; color photos.

351. Bessette, Alan, and Walter J. Sundberg. **Mushrooms: A Quick Reference Guide to Mushrooms of North America.** New York: Macmillan, 1987. 173p. (Macmillan Field Guides). ISBN 0026152606pa.

     200 species; color photos; grouped by characteristics.

352. Courtenay, Booth, and Harold H. Burdsall, Jr. **A Field Guide to Mushrooms and Their Relatives.** New York: Van Nostrand Reinhold, 1982. 144p. ISBN 0442231172; 0442231180pa.

     334 species; color photos; visual key; brief descriptions; and edibility. Gilled mushrooms are arranged by spore color, then size; other mushrooms are arranged by general physical characteristics.

353. Ellis, Martin B., and J. Pamela Elliot. **Fungi Without Gills (Hymenomycetes and Gasteromycetes): An Identification Guide.** New York: Chapman and Hall, 1990. 329p. ISBN 0412369702.

     974 species; key; line drawings; technical language.

354. Farr, Marie Leonore. **How to Know the True Slime Molds.** Dubuque, IA: W. C. Brown, 1981. 132p. $25.80. ISBN 0697047792pa.

     Introductory student's guide to common species; key throughout; black-and-white illustrations.

355. Fischer, David W., and Alan E. Bessette. **Edible Wild Mushrooms of North America: A Field-to-Kitchen Guide.** Austin: University of Texas Press, 1992. 288p. $40.00; $22.95pa. ISBN 0292720793; 0292720807pa.

99 species of edible mushrooms and 19 poisonous mushrooms; color photos; recipes.

356. Katsaros, Peter. **Familiar Mushrooms of North America**. New York: Knopf, 1990. 192p. (Audubon Society Pocket Guide Series). $9.00pa. ISBN 0679729844pa.

   80 species; color photos; includes description and information on edibility, similar species, habitat, and spore prints.

357. Katsaros, Peter. **Illustrated Guide to Common Slime Molds**. Eureka, CA: Mad River, 1989. 66p. $16.95pa. ISBN 0916422720pa.

   64 species; key; color illustrations; may need pocket lens to identify.

358. Kibby, Geoffrey. **Mushrooms and Other Fungi**. New York, NY: Smithmark, 1992. 92p. (American Nature Guides). $9.95. ISBN 083176970X.

   400 species; color illustrations.

359. Largent, David Lee. **How to Identify Mushrooms to Genus I: Macroscopic Features**. 2nd ed. Eureka, CA: Mad River, 1977. 86p. ISBN 0916422003pa.

   This student guide includes 74 North American genera; key; black-and-white illustrations. The bulk of this book consists of descriptions and illustrations of macroscopic features of mushrooms, which are then organized into an illustrated identification table. There is also a dichotomous key to genera. Illustrated by Sharon Hadley.

360. Largent, David Lee, and Harry D. Thiers. **How to Identify Mushrooms to Genus II: Field Identification of Genera**. 2nd ed. Eureka, CA: Mad River, 1977. 32p. ISBN 0916422089pa.

   74 North American genera; no illustrations. This volume, written for students, describes each genera of North American mushrooms using features observable in the field.

361. Lawrence, Eleanor, and Sue Harniess. **An Instant Guide to Mushrooms and Other Fungi: The Most Familiar Species of North American Mushrooms and Fungi Described and Illustrated in Full Color**. New York: Crescent, 1991. 124p. (Instant Guide Series). ISBN 0517691159.

   100 species; color illustrations; checklist.

362. Lincoff, Gary. **The Audubon Society Field Guide to North American Mushrooms**. New York: Knopf, 1981. 926p. (National Audubon Society Field Guide Series). $19.00 (flexi). ISBN 0394519922 (flexi).

    703 species covered in detail, many others briefly; color photos; spore print chart.

363. McIlvaine, Charles, and Robert K. Macadam. **One Thousand American Fungi: Toadstools, Mushrooms, Fungi, Edible and Poisonous with Full Botanic Descriptions**. Rev. ed. New York: Dover, 1973. 729p. $12.95pa. ISBN 0486227820pa.

    1,000 species; line drawings; natural history; edibility. Covers many species not found in other guides. Updated nomenclature. Reprint of the 1902 edition.

364. McKnight, Kent H., and Vera B. McKnight. **A Field Guide to Mushrooms, North America**. Boston: Houghton Mifflin, 1987. 429p. (Peterson Field Guide Series, No. 34). $18.00pa. ISBN 0395421012; 0395910900pa.

    1,000 species, 500 illustrated; color illustrations and line drawings; some recipes. Illustrated by Vera B. McKnight.

365. Menser, Gary P. **Hallucinogenic and Poisonous Mushroom Field Guide**. Berkeley, CA: Ronin, 1996. 124p. $14.95pa. ISBN 0914171895pa.

    24 species of hallucinogenic and 8 species of poisonous mushrooms; key; black-and-white illustrations and color photos. A straightforward field guide covering the spore print, season, range, and habitat of 32 species of mushrooms from the western United States, plus extensive general information on mycology and mushroom identification; contains no information on the ingestion of hallucinogenics. Originally published in 1977; reprinted in 1984 as the *Magic Mushroom Handbook*. Illustrated by Michael B. Smith.

366. Number not used.

367. Pacioni, Giovanni. **Simon and Schuster's Guide to Mushrooms**. New York: Simon and Schuster, 1981. 511p. ISBN 0671428489; 0671427989; 0671428497pa.

    420 species from North America and Europe; simple key; color photos. Translation of *Funghi*. Also published as *The Macdonald Encyclopedia of Mushrooms and Toadstools*, and in Spain as *Guía de Hongos*.

❏ Shuttleworth, Floyd S., and Herbert S. Zim. **Mushrooms and Other Non-Flowering Plants**. See entry 402.

368. Smith, Alexander Hanchett, and Nancy Smith Weber. **The Mushroom Hunter's Field Guide**. 2nd rev. ed. Ann Arbor: Co-published by University of Michigan Press and Thunder Bay Press, 1996. 316p. $24.95. ISBN 0472856103 (University of Michigan Press); 1882376242 (Thunder Bay Press).

300 species; keys; color photos; emphasis on edible species of the northern U.S.

369. Stamets, Paul. **The Field Guide to Psilocybin Mushrooms of the World**. Berkeley, CA: Ten Speed, 1996. 256p. $24.95pa. ISBN 0898158397pa.

150 species; key to genus; color photos; includes description, habitat, notes. This is a standard field guide in its format. The descriptions are similar to those in every other mushroom field guide, but the guide includes some information that may be illegal or dangerous to act upon.

370. Stephenson, Steven L., and Henry Stempen. **Myxomycetes: A Handbook of Slime Molds**. Portland, OR: Timber, 1994. 200p. $34.95. ISBN 0881922773.

Manual/field guide covering 175 common species; keys requiring hand lens or microscope; color and black-and-white illustrations. Much background information on slime molds.

371. Stuntz, D. E. **How to Identify Mushrooms to Genus IV: Keys to Families and Genera**. 2nd ed. Eureka, CA: Mad River, 1977. 94p. ISBN 0916422100pa.

Keys to family or genus; no illustrations. One key uses macroscopic features, the second uses microscopic features, and the third uses a combination of macroscopic and microscopic features. A companion to Largent's earlier volumes in this series (see entries 359 and 360).

372. Tekiela, Stan, and Karen Shanberg. **Start Mushrooming: The Easiest Way to Start Collecting 6 Edible Mushrooms**. Cambridge, MN: Adventure, 1993. 126p. $7.95pa. ISBN 0934860963pa.

6 species; black-and-white illustrations and color photos; 17 recipes. A beginner's guide to edible mushrooms. Covers morel, oyster, shaggy mane, sulfur shelf, giant puffball, and hen-of-the-woods mushrooms. Illustrated by Laverne Dunsmore.

Eastern North America

373. Bessette, Alan. **Mushrooms of the Adirondacks: A Field Guide**. Utica, NY: North Country, 1988. 145p. $12.95pa. ISBN 0932052649pa.

    140 species; key; color photos are separate from the text; species accounts include description, edibility, similar species; also includes 7 recipes.

374. Bessette, Alan, Arleen R. Bessette, and David W. Fischer. **Mushrooms of Northeastern North America**. Syracuse, NY: Syracuse University Press, 1996. 582p. $95.00; $29.95pa. ISBN 0815627076; 0815603886pa.

    1,500 species; keys, including color key to major groups and dichotomous keys to species; color photos; detailed descriptions and edibility. Covers the U.S. and Canada west to the Dakotas and south to Tennessee and North Carolina. A comprehensive, massive book not conducive to field work, but written in non-technical language.

375. Horn, Bruce, Richard Kay, and Dean Abel. **A Guide to Kansas Mushrooms**. Lawrence, KS: University Press of Kansas, 1993. 297p. (Kansas Nature Guides). $19.95pa; $29.95. ISBN 0700605711pa; 0700605703.

    150 species; keys; color photos; extensive background. Descriptions include similar species, habitat, edibility.

376. Huffman, D. M., L. H. Tiffany, and G. Knaphus. **Mushrooms and Other Fungi of the Midcontinental United States**. Ames, IA: Iowa State University Press, 1989. 326p. $24.95pa. ISBN 0813811686pa.

    200 species; color illustrations.

377. Kavasch, E. Barrie. **Guide to Eastern Mushrooms**. Killingworth, CT: Hancock House, 1982. 48p. (Northeast Color Series). $4.95pa. ISBN 0888390912.

    72 species; color photos; includes description, range, and edibility.

378. Lebrun, Denis, and Anne-Marie Guerineau. **Champignons du Québec et de l'Est du Canada**. Québec: Nuit blanche, 1988. 288p. ISBN 2921053004.

    105 species; color photos; descriptions of cap, stem, spores, etc. Includes notes on edibility and taste. First published as *Champignons du Québec*.

379. Metzler, Susan, and Van Metzler. **Texas Mushrooms: A Field Guide**. Austin: University of Texas Press, 1992. 304p. (A Corrie Herring Hooks Book). $39.95; $19.95pa. ISBN 0292751257; 0292751265pa.

    200 species; visual guide to family; color photos; recipes. Also includes tables listing species by spore print color.

380. Weber, Nancy S., and Alexander H. Smith. **A Field Guide to Southern Mushrooms**. Ann Arbor: University of Michigan Press, 1985. 280p. $24.95. ISBN 0472856154.

    241 species; keys; color photos.

Western North America

381. Arora, David. **All That the Rain Promises and More . . . : A Hip Pocket Guide to Western Mushrooms**. Berkeley, CA: Ten Speed, 1991. 261p. $17.95pa. ISBN 0060391103pa.

    200 edible and poisonous species; quick key; color photos; checklist; arranged by family. Includes interesting, often humorous, tidbits of information.

382. Evenson, Vera S. **Mushrooms of Colorado and the Southern Rocky Mountains**. Englewood, CO: Westcliffe, 1997. 224p. $25.00pa. ISBN 1565791924pa.

    170 species; keys; color photos; extensive background. Descriptions include details of fruiting body and spores, edibility, and general notes.

383. McKenny, Margaret, and Daniel E. Stuntz. **The New Savory Wild Mushroom**. 3rd ed. Seattle, WA: University of Washington Press, 1987. 249p. $19.95pa. ISBN 029596491X; 0295964804pa.

    156 species; black-and-white and some color photos; covers the Pacific Northwest. Revised edition of *The Savory Wild Mushroom*.

384. Miller, Millie. **Chanterelle: A Rocky Mountain Mushroom Book**. Boulder, CO: Johnson, 1986. 40 leaves. (Millie and Cyndi's Pocket Nature Guides). $5.95pa. ISBN 0933472978pa.

    40 species; color illustrations; pocket size.

385. Orr, Robert Thomas, and Dorothy B. Orr. **Mushrooms of Western North America**. Berkeley, CA: University of California Press, 1979. 293p. (California Natural History Guides, No. 42). $45.00; $12.95pa. ISBN 0520036565; 0520036603pa.

300 species illustrated, another 123 described; keys; color photos, some line drawings. Includes abbreviations of the names of taxonomic authors and a list of mycological collections in the west. Illustrated by Jacqueline Schonewald and Paul Verger.

386. Parker, Harriette. **Alaska's Mushrooms: A Practical Guide**. Anchorage: Alaska Northwest, 1994. 92p. (Alaska Pocket Guide). $12.95pa. ISBN 0882404539pa.

    35 species; color photos; multiple common names; natural history; edibility when known; and 8 recipes.

387. Schalkwyk, Helene M. E. **Mushrooms of Northwest North America**. Edmonton, Alta.: Lone Pine, 1994. 414p. $19.95pa. ISBN 1551050463pa.

    550 species; color illustrations; illustrations with facing brief descriptions followed by more extensive species accounts. Also published as *Mushrooms of Western Canada*.

388. Smith, Alexander Hanchett. **A Field Guide to Western Mushrooms**. Ann Arbor: University of Michigan Press, 1975. 280p. ISBN 0472855999.

    200 species; keys; color photos; includes description, edibility, distribution, and some microscopic traits.

389. States, Jack S. **Mushrooms and Truffles of the Southwest**. Tucson: University of Arizona Press, 1990. 232p. $15.00pa. ISBN 0816511826pa; 0816511624.

    156 species illustrated, another 155 briefly described; keys; color photos. A unique feature is that it includes cross-references to other standard field guides to aid in identification.

390. Thiers, Harry Delbert. **California Mushrooms: A Field Guide to the Boletes**. New York: Hafner, 1974. 261p. ISBN 0028534107.

    200 species; keys; field manual. Includes microfiche card with color photos of 54 species (to be used in field by examining with field lens).

391. Tylutki, Edmund E. **Mushrooms of Idaho and the Pacific Northwest, Volume One: Discomycetes: Morels, False Morels, Fairy Cups, Saddle Fungi, Earth Tongues, Truffles and Related Fungi (Ascomycetes-Discomycetes)**. 2nd rev. ed. Moscow, ID: University of Idaho Press, 1993. 133p. (A Northwest Naturalist Book). ISBN 0893010626pa.

    95 species; black-and-white photos; descriptions include fruiting bodies, microscopic traits, habitat, and remarks. Originally

conceived as a five-volume set, but apparently only volumes 1 and 2 were published.

392. Tylutki, Edmund E. **Mushrooms of Idaho and the Pacific Northwest, Volume Two: Non-Gilled Hymenomycetes**. Moscow, ID: University Press of Idaho, 1987. 232p. (A Northwest Naturalist Book). ISBN 0893010979pa.

    354 species of boletes, coral fungi, chanterelles, polypores, and spine fungi; keys; black-and-white and color photos.

Non-Flowering Plants

The field guides covered in this chapter identify algae, ferns, mosses, liverworts, lichens, and other non-flowering plants. Mushrooms, which belong to an entirely different kingdom, are covered in Chapter 5. The non-flowering plant field guides are generally similar to other plant guides, though guides to some of the groups may be more technical than most flower guides.

Other non-flowering plants are found in the guides in Chapter 2, "Flora and Fauna" and Chapter 3, "Plants." In addition, many flower field guides include a few conspicuous non-flowering plant species, and algae are often found in guides to aquatic life.

NORTH AMERICA

393. Cody, William J., and Donald M. Britton. **Ferns and Fern Allies of Canada**. Ottawa: Research Branch, Agriculture Canada, 1989. 430p. (Publication, No. 1829). $46.20. ISBN 0660131021.

    Comprehensive; field manual; keys; line drawings; maps. Published in French as *Les Fougeres et les Plantes Alliées du Canada*.

394. Conard, Henry S., and Paul L. Redfearn, Jr. **How to Know the Mosses and Liverworts**. 2nd ed., rev. Dubuque, IA: W. C. Brown, 1979. 302p. (Pictured Key Nature Series). $25.80pa. ISBN 0697047695; 0697047687pa.

    Common species; keys; line drawings. Previously published as *How to Know the Mosses*.

395. Dawson, Elmer Yale, and Isabella A. Abbott. **How to Know the Seaweeds.** 2nd ed. Dubuque, IA: W. C. Brown, 1978. 141p. (Pictured Key Nature Series). $29.95pa. ISBN 0697048950; 0697048926pa.

     240 species in 175 genera; keys; line drawings; background information.

396. Hale, Mason E., Jr. **How to Know the Lichens.** 2nd ed. Dubuque, IA: W. C. Brown, 1979. 246p. (Pictured Key Nature Series). $25.80pa. ISBN 0697047628; 0697047636pa.

     427 species; keys; line drawings and black-and-white photos. Excludes crustose forms.

397. Lellinger, David B. **A Field Manual of the Ferns and Fern-Allies of the United States and Canada.** Washington, DC: Smithsonian Institution Press, 1985. 389p. $29.95pa. ISBN 0874746027; 0874746035pa.

     406 species; key; color photos.

398. McQueen, Cyrus B. **Field Guide to the Peat Mosses of Boreal North America.** Hanover, NH: University Press of New England, 1990. 138p. $25.00. ISBN 087451522Xpa.

     26 common species described and illustrated; key to about 50 species; color photos.

399. Mickel, John T. **How to Know the Ferns and Fern Allies.** Dubuque, IA: W. C. Brown, 1979. 229p. (Pictured Key Nature Series). $25.80pa. ISBN 0697047709; 0697047717pa.

     400 species; keys; line drawings; maps. Illustrated by Edgar M. Paulton.

400. Prescott, G. W. **How to Know the Aquatic Plants.** 2nd ed. Dubuque, IA: W. C. Brown, 1980. 158p. (Pictured Key Nature Series). $25.80pa. ISBN 0697047741; 069704775Xpa.

     165 genera; keys to genus or common species; line drawings.

401. Prescott, G. W. **How to Know the Freshwater Algae.** 3rd ed. Dubuque, IA: W. C. Brown, 1978. 293p. (Pictured Key Nature Series). $25.80pa. ISBN 0697047555; 0697047547pa.

     558 genera; key to genus; line drawings.

402. Shuttleworth, Floyd S., and Herbert S. Zim. **Mushrooms and Other Non-Flowering Plants.** New York: Golden, 1987. 160p. (Golden Guide). ISBN 0307240142pa.

     450 common species including algae, fungi, lichens, mosses, liverworts, ferns, and gymnosperms; color illustrations. About

half of the book consists of fungi. Revised edition of *Non-Flowering Plants*. Illustrated by Dorothea and Sy Barlowe.

Eastern North America

403. Cobb, Boughton. **A Field Guide to the Ferns, and Their Related Families of Northeastern and Central North America with a Section on Species Also Found in the British Isles and Western Europe**. Boston: Houghton Mifflin, 1963. 281p. (Peterson Field Guide Series, No. 10). $18.00pa. ISBN 0395075602; 039597123pa.

    100 species; key; line drawings; checklists. Illustrated by Laura Louise Foster.

404. Crum, Howard Alvin. **Liverworts and Hornworts of Southern Michigan**. Ann Arbor: University of Michigan Herbarium, 1991. 233p. $18.00. ISBN 0962073318.

    75 species; keys; line drawings, black-and-white photos; field manual.

405. Dawes, Clinton J. **Marine Algae of the West Coast of Florida**. Coral Gables, FL: University of Miami Press, 1974. 201p. $18.95pa. ISBN 087024258Xpa.

    Comprehensive field manual; key; line drawings and black-and-white photos.

406. Glime, Janice M. **The Elfin World of Mosses and Liverworts of Michigan's Upper Peninsula and Isle Royale**. Houghton, MI: Isle Royale Natural History Association, 1993. 148p. $15.95pa. ISBN 0935289046pa.

    200 species; color photos; arranged by habitat. Illustrated by Marshall L. Strong.

407. Hallowell, Anne C., and Barbara G. Hallowell. **Fern Finder: A Guide to Native Ferns of Northeastern and Central North America**. Berkeley, CA: Nature Study Guild, 1981. 59p. $3.00pa. ISBN 0912550112pa.

    87 species; key throughout; black-and-white illustrations; maps; pocket sized. Illustrated by Anne C. Hallowell.

408. Hillson, Charles James. **Seaweeds: A Color-Coded, Illustrated Guide to Common Marine Plants of the East Coast of the United States**. University Park: Pennsylvania State University

Press, 1977. 194p. (Keystone Books). $30.00; $14.95pa. ISBN 0271012390; 0271012471pa.

100 species; black-and-white illustrations; arranged by color; also includes a few seagrasses.

409.   Lamoureux, Gisele, ed. **Fougères, Prèles et Lycopodes**. Saint-Henri-de-Levis, Québec: Fleurbec, 1993. 511p. (Guide d'Identification Fleurbec). ISBN 2920174134.

55 species of ferns, horsetails, and club mosses; color photos; maps; natural history and background information. In French.

410.   Lamoureaux, Gisele, ed. **Plantes Sauvages du Bord de la Mer**. Saint-Augustin, Québec: Fleurbec, 1985. 286p. (Guide d'Identification Fleurbec). ISBN 2920174088.

56 species of algae; color photos; maps; natural history; emphasis on seaweeds of the Atlantic Coast of Québec, but includes distribution throughout North America. In French.

411.   Littler, Diane Scullion, Mark M. Littler, Katina E. Bucher, and James N. Norris. **Marine Plants of the Caribbean: A Field Guide from Florida to Brazil**. Washington, DC: Smithsonian Institution Press, 1989. 263p. $24.95pa. ISBN 0874746078pa.

204 species of algae, 5 species of seagrass; color photos.

412.   Reese, William Dean. **Mosses of the Gulf South: From the Rio Grande to the Apalachicola**. Baton Rouge: Louisiana State University Press, 1984. 252p. $35.00. ISBN 0807111104.

200 species; key.

413.   Snyder, Lloyd H., Jr., and James G. Bruce. **Field Guide to the Ferns and Other Pteridophytes of Georgia**. Athens: University of Georgia Press, 1986. 270p. $15.00pa. ISBN 0820308382; 0820308471pa.

120 species; key; line drawings. Illustrations by Joseph H. Pyron from McVaugh's *Ferns of Georgia*.

414.   Tryon, Rolla Milton. **Ferns of Minnesota**. Minneapolis: University of Minnesota Press, 1980. 165p. $14.95pa. ISBN 0816609322; 0816609357pa.

100 species; key; black-and-white illustrations, some color photos. Earlier edition titled *Ferns and Fern Allies of Minnesota*. Illustrated by Wilma Monserud.

415.   Wherry, Edgar Theodore. **The Fern Guide, Northeastern and Midland United States and Adjacent Canada**. New York: Dover, 1995. 318p. $8.95pa. ISBN 0486284964pa.

Over 130 species; key; line drawings. Includes authors of scientific names, meaning of epithets, and other technical material. Reprint of the Doubleday Nature Guides Series, 9. Illustrated by James C. W. Chen.

416.   Wherry, Edgar Theodore. **The Southern Fern Guide, Southeastern and South-Midland United States**. Corrected ed., with nomenclatural changes. Bronx: New York Chapter of the American Fern Society, 1978. 349p.

200 species; key; line drawings. Reprint of the Doubleday Nature Guide Series. Illustrated by James C. W. Chen and Keith C. Y. Chen.

417.   Wiley, Farida A. **Ferns of Northeastern United States; Illustrations and Descriptions of All Known Species in the New England and Middle Atlantic States**. New York: Dover, 1964. 108p. $3.95pa. ISBN 0486229467pa.

65 species; line drawings. Reprint of the 2nd ed. (1948).

Western North America

418.   Corbridge, James N., and William A. Weber. **Rocky Mountain Lichen Primer**. Niwot, CO: University Press of Colorado, 1998. 47p. $19.95pa. ISBN 0870814907pa.

72 common species; color photos; descriptions include general notes and habitat. Also includes background information.

419.   Grillos, Steve J. **Ferns and Fern Allies of California**. Berkeley, CA: University of California Press, 1966. 104p. (California Natural History Guides, No. 16). $9.95pa. ISBN 0520030915; 0520005198pa.

86 species; keys; line drawings. Illustrated by Rita Whitmore.

420.   Hale, Mason E., Jr., and Mariette Cole. **Lichens of California**. Berkeley, CA: University of California Press, 1988. 254p. (California Natural History Guides, No. 54). $16.95pa. ISBN 0520057120; 0520057139pa.

350 species; keys; black-and-white illustrations; maps.

421.   Keator, Glenn, and Ruth M. Heady. **Pacific Coast Fern Finder**. Berkeley, CA: Nature Study Guild, 1981. $3.00pa. ISBN 0912550139pa.

175 species; key throughout; black-and-white illustrations. Pocket-sized pamphlet.

422. Magruder, William H., and Jeffrey W. Hunt. **Seaweeds of Hawaii: A Photographic Identification Guide**. Honolulu: Oriental, 1979. 116p. ISBN 0932596126.
      150 species; color photographs.

423. McCune, Bruce, and Linda Geiser. **Macrolichens of the Pacific Northwest**. Corvallis: Oregon State University Press, 1997. 386p. $25.95pa. ISBN 0870713949pa.
      458 species; keys; color photos and black-and-white illustrations for 200 species; detailed, technical descriptions include habitat, range, and substrate. Background information, including air quality sensitivities for 100 species. Illustrated by Alexander Mikulin.

424. McCune, Bruce, and Trevor Goward. **Macrolichens of the Northern Rocky Mountains**. Eureka, CA: Mad River, 1995. 208p. ISBN 0916422828pa.
      150 species; keys; black-and-white illustrations; checklist. Spiralbound. Illustrated by Daphne F. Stone.

425. O'Clair, Rita M. **Southeast Alaska's Rocky Shores: Seaweeds and Lichens**. Auke Bay, AK: Plant Press, 1996. 151p.
      83 algae, 30 lichens, 1 moss, 2 seagrasses; black-and-white illustrations; includes recipes and notes.

426. Valier, K. **Ferns of Hawaii**. Honolulu: University of Hawaii Press, 1995. 112p. $14.95pa. ISBN 0824816404pa.
      Covers 60 of Hawaii's 200 species, plus a few fern allies; black-and-white and color photos; English and Hawaiian common names. Includes information on use of ferns by native Hawaiians.

427. Vitt, Dale H., Janet E. Marsh, and Robin B. Bovey. **Mosses, Lichens and Ferns of Northwest North America: A Photographic Field Guide**. Edmonton, Alta.: Lone Pine, 1988. 296p. $24.95pa. ISBN 0919433413pa.
      370 common species; keys; color photos; maps.

| Chapter 7 | # Trees and Shrubs |

Trees, shrubs, and other woody plants, such as vines, are found in this chapter. This includes both coniferous and deciduous trees, plus flowering shrubs. Some guides to cultivated or non-native trees and shrubs are included, but the majority of the guides are to wild-growing plants. Since trees are of economic importance, there are many tree field guides that discuss the use and economic value of tree species. A number of these practical guides have been published by various state extension services, agricultural departments, or university presses.

Many tree guides illustrate only leaves, but others include illustrations and descriptions of leaves, bark, fruit or nuts, twigs, buds, and the general shape of the tree. Keys are common in tree guides, especially winter keys to trees without leaves.

Trees are also commonly found in the field guides in Chapter 2, "Flora and Fauna," and Chapter 3, "Plants."

NORTH AMERICA

428. Brockman, C. Frank. **Trees of North America: A Field Guide to the Major Native and Introduced Species North of Mexico**. Rev. ed. New York: Golden, 1986. 280p. (A Golden Field Guide). $11.95pa. ISBN 0307136582pa.

594 native species plus 36 introduced; color illustrations; maps; illustrations include silhouette, bark, berry or fruit, flower, and twig as well as leaf. Text and illustrations integrated. Illustrated by Rebecca Merrilees.

429. Constantine, Albert. **Know Your Woods: A Complete Guide to Trees, Woods, and Veneers**. Rev. ed. New York: Scribner, 1987. 360p. $14.00pa. ISBN 0684187787pa.

    300 types of wood; black-and-white photos and illustrations; maps. The types of woods covered include commercially available species worldwide (such as mahogany), plus most North American trees. For the non-North American species, the wood is described and the distribution and multiple common names are given, but there are no illustrations of living trees. North American trees, however, are illustrated and have range maps.

430. Edlin, Herbert L. **What Wood Is That? A Manual of Wood Identification**. New York: Viking, 1969. 160p. $29.95pa. ISBN 0670759074pa.

    40 species of trees; keys; black-and-white illustrations; includes fold-out section with actual wood samples; worldwide. Extensive introductory material on lumbering, wood preparation, and other topics. Descriptions include original range, use, descriptions of wood and tree, and multiple common names.

431. Elias, Thomas S. **Field Guide to North American Trees**. Rev. ed. Danbury, CT: Grolier Book Clubs, 1989. 948p. $32.95. ISBN 155654037Xpa; 1556540493.

    750 species; visual and dichotomous keys; line drawings. Includes photos of bark types. Originally published as *The Complete Trees of North America*. Cover title: *Grolier's Field Guide to North American Trees*.

432. Farjon, Aljos, Jorge A. Perez de Rosa, and Brian T. Styles. **A Field Guide to the Pines of Mexico and Central America**. Richmond, Surrey, U.K.: Royal Botanic Gardens, 1997. 147p. ISBN 1900347369pa.

    43 species; black-and-white illustrations; maps. Also available in Spanish version as *Guía de Campo de los Pinos de Mexico y America Central*.

433. Farrar, John Laird. **Trees of the Northern United States and Canada**. Ames, IA: Iowa State University Press, 1995. 502p. $39.95. ISBN 081382740X.

    300 species; keys; color photos and line drawings; maps; arranged by leaf shape and arrangement; oversized. For students, foresters, and others. Common names in English and French. Published in Canada as *Trees in Canada*.

434. Henderson, Andrew, Gloria Galeano, and Rodrigo Bernal. **Field Guide to the Palms of the Americas**. Princeton, NJ: Princeton University Press, 1995. 376p. $29.50pa. ISBN 0691085374; 0691016003pa.
550 species; keys; line drawings; maps; technical language.

435. Mitchell, Alan F. **Spotter's Guide to Trees of North America**. 1st American ed. New York: Mayflower, 1979. 64p. $12.95; $4.95pa. ISBN 0831788186; 0831788194pa.
87 species; color illustration; background information. Also published in Spanish by Juvental.

436. Mitchell, Alan F. **Trees**. New York: Gallery, 1990. 224p. (American Nature Guides). ISBN 0831769602.
115 species; color illustrations; range maps with winter silhouettes in separate section; includes chapter on selection and cultivation of trees. Illustrated by David More.

437. Mohlenbrock, Robert H., and John W. Thieret. **Trees: A Quick Reference Guide to Trees of North America**. New York: Macmillan, 1987. 155p. (Macmillan Field Guides). ISBN 0025854607pa.
232 species; color illustrations; maps. Illustrated by Howard S. Friedman.

438. Petrides, George A. **Peterson First Guide to Trees**. Boston: Houghton Mifflin, 1993. 128p. $5.95pa. ISBN 0395911834pa.
243 species; color illustrations. Illustrated by Olivia Petrides and Janet Wehr.

439. Preston, Richard Joseph, Jr. **North American Trees: Exclusive of Mexico and Tropical Florida**. 4th ed. Ames, IA: Iowa State University Press, 1989. 407p. $41.95; $22.95pa. ISBN 0813811716; 0813811724pa.
570 species; key; black-and-white drawings; maps.

❑ Reader's Digest Editors. **North American Wildlife: Trees and Nonflowering Plants**. See entry 197.

440. Rushforth, Keith. **The Pocket Guide to Trees**. New York: Simon and Schuster, 1981. 215p. $7.95. ISBN 0671255142.
450 species; color illustrations; covers North America and Europe. Also published as *The Mitchell Beazley Pocket Guide to Trees*, and in Spanish as *Guía de los Arboles*.

441.   Zim, Herbert Spencer, and Alexander C. Martin. **Trees: A Guide to Familiar American Trees: 143 Species in Color**. New York: Golden, 1991. 160p. (Golden Guide). $5.95pa. ISBN 0307640566; 0307240568pa.

   143 species; color illustrations; maps. Illustrated by Dot and Sy Barlowe.

Eastern North America

442.   Baldwin, Henry Ives. **Forest Leaves: How to Identify Trees and Shrubs of Northern New England**. 2nd ed. Portsmouth, NH: P. R. Randall, 1993. 255p. $12.95pa. ISBN 0914339435pa.

   123 species; line drawings of leaves only, with descriptions of bark and tree shapes. Illustrated by Gunnar I. Baldwin and Priscilla Kunhardt.

443.   Barnes, Burton V., and Warren H. Wagner, Jr. **Michigan Trees: A Guide to the Trees of Michigan and the Great Lakes Region**. Ann Arbor: University of Michigan Press, 1981. 383p. $16.95pa. ISBN 0472080172; 0472080180pa.

   240 species of trees and 100 common shrubs; summer and winter keys; line drawings. Revised edition of Charles Herbert Otis's *Michigan Trees*.

444.   Bell, C. Ritchie, and Anne H. Lindsey. **Fall Color and Woodland Harvests: A Guide to the More Colorful Fall Leaves and Fruits of the Eastern Forests**. Chapel Hill, NC: Laurel Hill, 1990. 184p. $16.75. ISBN 096086881X.

   100 species; key; color photos; maps. Covers fall leaves, berries, and seeds.

445.   Bell, C. Ritchie, and Anne H. Lindsey. **Fall Color Finder: A Pocket Guide to the More Colorful Trees of Eastern North America**. Chapel Hill, NC: Laurel Hill, 1991. 63p. $4.95pa. ISBN 0960868828pa.

   50 species; key throughout; color photos; maps; includes leaf shape/color index. Pocket sized.

446.   Blouin, Glen. **Weeds of the Woods: Small Trees and Shrubs of the Eastern Forest**. Fredericton, N.B.: Goose Lane, 1992. 125p. $14.95. ISBN 0864921276.

   20 species; color photos; natural history. Most species are illustrated by twigs, bark, leaves, flowers, and fruit. Covers eastern Canada and the northeastern U.S. Originally published in 1984 as

Weeds of the Woods: Some Small Trees and Shrubs of New Brunswick and *Arbustes des Bois: La Flore Meconnue du Nouveau-Brunswick.*

447. Brown, Claud L., and L. Katerine Kirkman. **Trees of Georgia and Adjacent States**. Portland, OR: Timber, 1990. 292p. $39.95. ISBN 0881921483.

    205 species; summer and winter keys; color photos are separate from the text; maps.

448. Choukas-Bradley, Melanie. **City of Trees: The Complete Field Guide to the Trees of Washington, D.C.** Rev. ed. Baltimore: Johns Hopkins University Press, 1987. 354p. $15.95pa. ISBN 0801833205pa.

    300 species; keys to genus; black-and-white illustrations and color photos; includes tree viewing sites and appendix with flowering periods. descriptions include native habitat, similar species, and locations found. The first edition, while containing the same information, is oversized and should not be considered a field guide.

449. Cope, Edward A. **Native and Cultivated Conifers of Northeastern North America: A Guide**. Ithaca, NY: Comstock, 1986. 231p. $45.00; $16.95pa. ISBN 080141721X; 0801493609pa.

    130 species; keys; black-and-white photos and line drawings of trees, needles, cones; includes list of over 2,700 cultivars.

450. Cox, Paul W., and Patty Leslie. **Texas Trees: A Friendly Guide**. San Antonio: Corona, 1988. 374p. $10.95pa; $18.95. ISBN 0931722675pa; 0931722667.

    175 species; pictorial leaf key; black-and-white illustrations; natural history and use. Illustrated by Gloria Merlot and Sara Harrison.

451. Dean, Blanche Evans. **Trees and Shrubs of the Southeast**. 3rd ed., rev. and expanded. Birmingham, AL: Birmingham Audubon Society, 1988. 264p. ISBN 0962077909.

    840 species; key to families; line drawings and black-and-white photos. Revised and expanded edition of *Trees and Shrubs in the Heart of Dixie.*

452. Duncan, Wilbur Howard, and Marion B. Duncan. **Trees of the Southeastern United States**. Athens: University of Georgia Press, 1988. 322p. (Wormsloe Foundation Publications, No. 18). $19.95pa. ISBN 0820309540; 0820314692pa.

    300 species; keys; color photos; maps.

453.   Dwelley, Marilyn J. **Trees and Shrubs of New England**. Camden, ME: Down East, 1980. 275p. ISBN 0892720646.

   330 species; color illustrations including leaves, buds, flowers, etc.

454.   Grimm, William Carey. **The Illustrated Book of Trees: With Keys for Summer and Winter Identification**. Harrisburg, PA: Stackpole, 1983. 493p. $22.95pa. ISBN 0811722201pa.

   250 species; winter and summer keys; line drawings; covers eastern U.S.; oversized. Illustrated by the author.

455.   Gupton, Oscar W., and Fred C. Swope. **Trees and Shrubs of Virginia**. Charlottesville: University Press of Virginia, 1981. 205p. $14.95. ISBN 0813908868.

   100 species illustrated with color photos, with another 85 species described.

456.   Harlow, William Morehouse. **Fruit Key and Twig Key to Trees and Shrubs: Fruit Key to Northeastern Trees, Twig Key to the Deciduous Woody Plants of Eastern North America**. New York: Dover, 1959. 56p. $3.95pa. ISBN 0486205118pa.

   Fruit key section covers 120 species; twig key section covers 160; black-and-white photos of fruits and twigs. Reprint of the first edition of *Fruit Key to Northeastern Trees*, and the fourth revised edition of *Twig Key to the Deciduous Woody Plants of Eastern North America*.

457.   Harlow, William Morehouse. **Trees of the Eastern and Central United States and Canada**. New York: Dover, 1957. 288p. $6.95pa. ISBN 0486203956.

   140 species; keys; numerous black-and-white photos of bark, twigs, silhouette, etc. First published in 1942 as *Trees of the Eastern United States and Canada*.

458.   Harrar, Ellwood S., and J. George Harrar. **Guide to Southern Trees**. 2nd ed. New York: Dover, 1962. 709p. $11.95pa. ISBN 0486209458pa.

   350 species; keys; line drawings; descriptions include leaves, flowers, and bark. Includes Florida Keys and western Texas.

459.   Hunter, Carl G. **Autumn Leaves and Winter Berries in Arkansas**. Little Rock, AR: Ozark Society Foundation, 1995. 52p. $12.95pa. ISBN 0912456205pa.

   45 tree species and 42 berries; color photos; brief descriptions. Pamphlet. Can be used as a supplement to the author's *Trees, Shrubs and Vines of Arkansas* (see entry 460).

460. Hunter, Carl G. **Trees, Shrubs and Vines of Arkansas**. Little Rock, AR: Ozark Society Foundation, 1995. 207p. $29.50; $22.50pa. ISBN 0912456183; 0912456191pa.

   484 species illustrated with color photos, 116 species with descriptions only; arranged by families.

461. Knobel, Edward. **Identify Trees and Shrubs by Their Leaves: A Guide to Trees and Shrubs Native to the Northeast**. New York: Dover, 1972. 47p. $4.95pa. ISBN 0486228967pa.

   214 species; key; black-and-white illustrations. First published in 1894 as *A Guide to Find the Names of All Wild-Growing Trees and Shrubs of New England by Their Leaves*.

462. Lanner, Ronald M. **Autumn Leaves: A Guide to the Fall Colors of the Northwoods**. Minocqua, WI: NorthWord, 1990. 181p. (Northwood Nature Guide Collection). $9.95pa. ISBN 1559710780pa.

   85 species of trees, including evergreens; color photos; covers northeastern North America. More a natural history than a field guide, but photos are useful for identifying trees in fall; includes information on animals associated with the area.

463. Lawrence, Gale. **Vermont Life's Guide to Fall Foliage**. 2nd ed. Montpelier, VT: Vermont Life, 1993. 49p. $4.95pa. ISBN 0936896256pa.

   15 species; black-and-white illustrations; arranged by leaf color; includes suggested fall color viewing trips. Pocket sized. Illustrated by Adelaide Murphy.

464. Leopold, Donald Joseph, William C. McComb, and Robert N. Muller. **Trees of the Central Hardwood Forests of North America: An Identification and Cultivation Guide**. Portland, OR: Timber, 1998. 469p. $49.95. ISBN 0881924067.

   188 native and 84 cultivated species; summer and winter keys; black-and-white and color photos; maps. Each part of the tree is described (bark, leaves, twigs, etc.), and wildlife value and cultivation information are included.

465. Little, Elbert L. **The Audubon Society Field Guide to North American Trees: Eastern Region**. New York: Knopf, 1980. 714p. (National Audubon Society Field Guide Series). $19.00 (flexi). ISBN 0394507606 (flexi).

   400 species; keys; color photos and line drawings.

466. Lonard, Robert I., James H. Everitt, and Frank W. Judd. **Woody Plants of the Lower Rio Grande Valley, Texas**. Austin: Texas Memorial Museum, University of Texas at Austin, 1991. 179p. (Miscellaneous Publications, No. 7).

202 species; keys; black-and-white illustrations. Illustrated by Norman A. Browne.

467. McCoy, Doyle. **Roadside Trees and Shrubs of Oklahoma**. Norman: University of Oklahoma Press, 1981. 116p. $12.95pa. ISBN 0806115564pa.

156 species; color photos.

468. Michaelson, M. **Firewood: A Woodcutter's Fieldguide to Trees in Summer and Winter**. Mankato, MN: Gabriel, 1978. 157p.

75 species; black-and-white photos and line drawings; maps; describes and illustrates trees in summer and winter, including bark, trees, fruits, etc.

469. Miller, Howard A., and H. E. Jaques. **How to Know the Trees**. 3rd ed. Dubuque, IA: W. C. Brown, 1978. 263p. (Pictured Key Nature Series). $25.80pa. ISBN 0697048977; 0697048969pa.

360 species; keys; black-and-white photos, line drawings; maps; includes winter key and list of rare and endangered species. Covers eastern U.S. Earlier editions by H. E. Jaques.

470. Muenscher, Walter Conrad Leopold. **Keys to Woody Plants**. 6th rev. ed. Ithaca, Cornell University Press, 1955. 108p. $11.95pa. ISBN 0801403073pa.

550 species; summer and winter keys; covers northeastern U.S.

471. Nelson, Gil. **The Shrubs and Woody Vines of Florida: A Reference and Field Guide**. Sarasota, FL: Pineapple, 1996. 391p. $30.95; $21.95pa. ISBN 1561641065; 1561641103pa.

550 species of native and naturalized species; color photos and line drawings. Companion to the author's *Trees of Florida*.

472. Nelson, Gil. **The Trees of Florida: A Reference and Field Guide**. Sarasota, FL: Pineapple, 1994. 338p. $19.95pa. ISBN 1561640530; 1561640557pa.

342 species; line drawings and color photos. Half of the book consists of natural history, with separate field guide section. Illustrated by R. Marvin Cook, Jr.

473. Petrides, George A. **A Field Guide to Eastern Trees: Eastern United States and Canada, Including the Midwest**. 1st ed., expanded. Boston: Houghton Mifflin, 1998. 424p. (Peterson Field Guide Series, No. 11). $19.00pa. ISBN 0395904552pa.

    455 species; color illustrations and photos; maps. Identification plates with color illustrations in the front, species accounts with maps and color photos in back. Revision of *A Field Guide to Trees and Shrubs* (see entry 474), omitting shrubs but adding 133 Florida species. Illustrated by Janet Wehr.

474. Petrides, George A. **A Field Guide to Trees and Shrubs: Field Marks of All Trees, Shrubs, and Woody Vines That Grow Wild in the Northeastern and North-Central United States and in Southeastern and South-Central Canada**. 2nd ed. Boston: Houghton Mifflin, 1972. 428p. (Peterson Field Guide Series, No. 11A). $16.95pa. ISBN 0395175798pa.

    646 species; keys; color illustrations and line drawings; west to the Great Plains but excludes the Southeast. Illustrated by the author and Roger Tory Peterson.

475. Powell, A. Michael. **Trees and Shrubs of the Trans-Pecos and Adjacent Areas**. Rev. ed. Austin: University of Texas Press, 1998. 464p. $75.00; $34.95pa. ISBN 0292765797; 0292765738pa.

    453 species; keys; line drawings. Rather large. Originally published as *Trees and Shrubs of Trans-Pecos Texas*.

476. Rathke, David M. **Minnesota Trees**. Rev. ed. St. Paul, MN: Minnesota Extension Service, University of Minnesota, College of Natural Resources, 1995. 94p. $10.75pa. ISBN 0962311618pa.

    150 species; keys, including visual shortcut guide; line drawings of leaf, twig, seed; checklist; natural history; and use. First edition published as *Minnesota's Forest Trees*.

477. Scurlock, J. Paul. **Native Trees and Shrubs of the Florida Keys: A Field Guide**. 3rd ed. Pittsburgh, PA: Laurel and Herbert, 1996. 220p. $37.95pa. ISBN 0614145856.

    170 species; visual keys; color photos; natural history.

478. Simpson, Benny J. **A Field Guide to Texas Trees**. Houston, TX: Gulf, 1992. 372p. (Texas Monthly Fieldguide Series). $18.95pa. ISBN 0877191131pa.

    222 species; color photos; maps; checklist. A very hefty field guide.

479. Smith, Norman Foster. **Trees of Michigan and the Upper Great Lakes**. Rev. 6th ed. Lansing, MI: Thunder Bay, 1995. 178p. $21.95pa. ISBN 1882376080pa.

   80 species; color photos; natural history. Each species is discussed in a two-page spread with photos of tree, bark, leaves; extensive description including range, natural history, use; also has box with identification characteristics. Revised edition of the 5th edition of *Michigan Trees Worth Knowing*.

480. Soper, James H., and Margaret L. Heimburger. **Shrubs of Ontario**. Toronto: Royal Ontario Museum, 1982. 495p. (Life Sciences Miscellaneous Publication of the Royal Ontario Museum). $35.00. ISBN 0888542836pa.

   Field manual; keys; line drawings; maps. Illustrated by Leslie A. Garay and Ronald A. With.

481. Swanson, Robert E. **A Field Guide to the Trees and Shrubs of the Southern Appalachians**. Baltimore: Johns Hopkins University Press, 1994. 399p. $18.95pa; $55.00. ISBN 0801845564pa; 0801845556.

   150 species; keys; line drawings. Illustrated by Frances R. Swanson.

482. Symonds, George Wellington Dillingham. **The Shrub Identification Book: The Visual Method for the Practical Identification of Shrubs, Including Woody Vines and Ground Covers**. New York: William Morrow, 1963. 379p. $19.50pa. ISBN 0688050409pa.

   300 species; pictorial keys (thorns, leaves, twigs, bark, and needles); black-and-white photos; little description. The keys take up about half the volume, with a separate "Main Pages" section featuring photos of silhouettes, bark, twigs, leaves, seeds; contains very little description. Eastern half of the U.S. and Canada. Oversized.

483. Symonds, George W. D. **The Tree Identification Book: A New Method for the Practical Identification and Recognition of Trees**. New York: M. Barrows, 1958. 272p. $17.95pa. ISBN 0688050395pa.

   200 species; pictorial keys (leaves, twigs, bark, and needles); black-and-white photos; little description. The keys take up about half the volume, with a separate "Main Pages" section featuring photos of silhouettes, bark, twigs, leaves, seeds; contains very little description. Eastern U.S. Oversized.

484. Taylor, Richard B., Jimmy Rutledge, and Joe G. Herrera. **A Field Guide to Common South Texas Shrubs**. Austin, TX: Texas Parks and Wildlife, Wildlife Division, 1997. 106p. $19.95pa. ISBN 1885696140pa.

44 species; color photos; divided into thorned and thornless species, then arranged by family. Descriptions include value to livestock and wildlife and crude protein values of vegetation.

485. Tiner, Ralph W. **Winter Guide to Woody Plants of Wetlands and Their Borders: Northeastern United States**. Leverett, MA: Institute for Wetland and Environmental Education and Research, 1997. 91p. $16.00pa.

270 species; keys; black-and-white illustrations; oversized; spiralbound. Each species description includes similar species, habitat, range, and wetland indicator status.

486. Vines, Robert A. **Trees of Central Texas**. Austin: University of Texas Press, 1984. 405p. $14.95pa. ISBN 0292780583pa.

186 species; black-and-white illustrations; key to Crataegus genus (rose family).

487. Vines, Robert A. **Trees of East Texas**. Austin: University of Texas Press, 1977. 538p. $17.95pa. ISBN 0292780168; 0292780176pa.

300 species; black-and-white illustrations; by family.

488. Vines, Robert A. **Trees of North Texas**. Austin: University of Texas Press, 1981. 466p. (Elma Dill Russell Spencer Foundation Series, No. 14). $24.95. ISBN 0292780184; 0292780206pa.

425 species; black-and-white illustrations; by family.

489. Watts, May Theilgaard. **Tree Finder: A Manual for the Identification of Trees by Their Leaves**. Berkeley, CA: Nature Study Guild, 1986. 58p. $3.00pa. ISBN 0912550015pa.

161 species; key throughout; line drawings; pocket sized. Covers eastern North America.

490. Watts, May Theilgaard, and Tom Watts. **Winter Tree Finder: A Manual for Identifying Deciduous Trees in Winter**. Berkeley, CA: Nature Study Guild, 1970. 58p. $3.00pa. ISBN 0912550031pa.

125 species; key throughout; black-and-white illustrations; maps. Pocket sized. Covers deciduous trees in eastern North America.

491. Wharton, Mary E., and Roger W. Barbour. **Trees and Shrubs of Kentucky**. 2nd ed. Lexington: University Press of Kentucky, 1994. 592p. (Kentucky Nature Studies, No. 4). $29.95. ISBN 081311294X.

     282 species; color and black-and-white photos separate from species accounts. Identification is by means of color photos arranged by general type (flowers, fruits, alternate leaves, etc.); separate species accounts include general description, range, and use.

492. Whitman, Ann, ed. **Familiar Trees of North America: Eastern Region**. New York: Knopf, 1986. 192p. (Audubon Society Pocket Guides). $9.00pa. ISBN 0394748514pa.

     78 species; color photos and tree silhouettes; covers area west to the Rocky Mountains.

Western North America

493. Arno, Stephen F. **Northwest Trees**. Seattle, WA: Mountaineers, 1977. 161p. $30.00; $12.95pa. ISBN 0916890554; 0916890503pa.

     35 species native to the Pacific Northwest; visual key; black-and-white illustrations. Illustrated by Ramona Hammerly.

494. Baerg, Harry J. **How to Know the Western Trees**. 2nd ed. Dubuque, IA: W. C. Brown, 1973. 179p. (Pictured Key Nature Series). ISBN 0697048012; 0697048004pa.

     383 species described, 247 illustrated with line drawings; key throughout; maps; no shrubs.

495. Bernstein, Art. **Native Trees of the Northwest: A Pocket Guide**. Grants Pass, OR: New Leaf, 1988. 119p. ISBN 0961752521pa.

     46 species; line illustrations and black-and-white photos; maps.

496. Bowers, Janice Emily. **Shrubs and Trees of the Southwest Deserts**. Tucson, AZ: Southwest Parks and Monuments Association, 1993. 140p. $9.95pa. ISBN 1877856347pa.

     115 species; black-and-white illustrations; arranged by color on colored pages; includes some cacti. Illustrated by Brian Wignall.

497. Brayshaw, T. Christopher. **Trees and Shrubs of British Columbia**. Vancouver, B.C.: University of British Columbia Press, 1996. 374p. (Royal British Columbia Museum Handbook). $24.95pa. ISBN 0774805641pa.

314 species plus subspecies and varieties; keys; black-and-white illustrations of leaves, flowers, fruits, and wood parts, plus some color photos; checklist. Description includes size, leaves, flowers, fruit, and range. A fairly technical handbook, but useful for amateurs. Illustrated by the author.

498. Carter, Jack L. **Trees and Shrubs of Colorado**. Boulder, CO: Johnson, 1988. 165p. $13.95pa. ISBN 0961994509pa.
    230 species; key throughout; black-and-white illustrations. Illustrated by Marjorie C. Leggitt.

499. Carter, Jack L. **Trees and Shrubs of New Mexico**. Boulder, CO: Mimbres, 1997. 534p. $29.95pa. ISBN 0965840409pa.
    430 species; keys; line drawings, maps. Includes cacti. A hefty, fairly technical work.

500. Crittenden, Mabel. **Trees of the West**. Blaine, WA: Hancock House, 1992. 220p. $14.95pa. ISBN 0888392633pa.
    160 species; keys; line drawings of silhouettes, leaves, twigs, seeds, flowers. Illustrated by the author and Jack Popovich.

501. Elmore, Francis Hapgood. **Shrubs and Trees of the Southwest Uplands**. Globe, AZ: Southwest Parks and Monuments Association, 1976. 214p. (Popular Series, Southwest Parks and Monuments Association, No. 19). $9.95pa. ISBN 091140841Xpa.
    168 species; key to evergreens; black-and-white illustrations, some color photos; colored pages arranged by habitat; includes natural history and use and both English and Spanish common names. Covers Arizona and New Mexico, plus southern Utah and Colorado. Illustrated by Jeanne R. Janish.

502. Kepler, Angela Kay. **Trees of Hawai'i**. Honolulu: University of Hawaii Press, 1991. 89p. (Kolowalu Book). $12.95pa. ISBN 0824813294pa.
    100 species; color photos; arranged by flower color.

503. Little, Elbert L. **The Audubon Society Field Guide to North American Trees: Western Region**. New York: Knopf, 1980. 640p. (National Audubon Society Field Guide Series). $19.00 (flexi). ISBN 0394507614 (flexi).
    350 species; keys; color photos and line drawings.

504. McMinn, Howard E., and Evelyn Maino. **An Illustrated Manual of Pacific Coast Trees**. 2nd ed. Berkeley, CA: University of California Press, 1980. 409p. $45.00; $16.95pa. ISBN 0520008464; 0520043642pa.

400 species; keys; line drawings. Originally published in 1937.

505. Metcalf, Woodbridge. **Native Trees of the San Francisco Bay Region**. Berkeley, CA: University of California Press, 1959. 72p. (California Natural History Guides, No. 4). $10.00pa. ISBN 0520008537pa.

52 species; black-and-white and color illustrations; silhouettes.

506. Metcalf, Woodbridge. **Introduced Trees of Central California**. Berkeley, CA: University of California Press, 1968. 159p. (California Natural History Guides, No. 27). $8.95pa. ISBN 0520015487pa.

200 species; line drawings; background information.

507. Parks, Catherine G., Evelyn L. Bull, and Torolf R. Torgersen. **Field Guide for the Identification of Snags and Logs in the Interior Columbia River Basin**. Portland, OR: U.S. Dept. of Agriculture, Forest Service, Pacific Northwest Research Station, 1997. 40p. (General Technical Report PNW, No. GTR 390).

10 coniferous and 3 deciduous species, plus 6 indications of fungal decay; color photos of dead trees with varying amounts of branches, cones, and bark attached. Includes detailed descriptions of bark, general form, branches and wood, and wildlife use. Companion to General Technical Report PNW No. GTR 391, *Trees and Logs Important to Wildlife in the Interior Columbia River Basin*.

508. Paruk, Jim. **Sierra Nevada Tree Identifier**. Yosemite, CA: Yosemite Association, 1996. 96p. $9.95pa. ISBN 0939666839pa.

44 native species; visual key; black-and-white illustrations; descriptions include habitat and range, similar species, and associated species. Appendixes list dominant tree types by altitude, common non-native species, and compare Ponderosa and Jeffrey pines.

509. Peterson, Peter Victor. **Native Trees of Southern California**. Berkeley, CA: University of California, 1966. 136p. (California Natural History Guides, No. 14). ISBN 0520030966; 0520010043pa.

100 species; key; line drawings, some color illustrations. Includes silhouettes. Illustrated by Rita Whitmore.

510. Peterson, Peter Victor, and P. Victor Peterson, Jr. **Native Trees of the Sierra Nevada**. Berkeley, CA: University of California, 1975. 147p. (California Natural History Guides, No. 36). ISBN 0520027361; 0520026667pa.

100 species; key; line drawings, some color illustrations. Includes silhouettes.

511. Petrides, George A. **A Field Guide to Western Trees: Western United States and Canada**. 1st ed., expanded. Boston: Houghton Mifflin, 1998. 428p. (Peterson Field Guide Series, No. 44). $19.00pa. ISBN 0395904544pa.

     387 species west of the Rockies and Black Hills; color illustrations and photos; short key; maps. Illustrated by Olivia Petrides.

512. Petrides, George A., and Olivia Petrides. **Trees of the California Sierra Nevada: A New and Simple Way to Identify and Enjoy Some of the World's Most Beautiful and Impressive Forest Trees in a Mountain Setting of Incomparable Majesty**. Williamston, MI: Explorer, 1996. 79p. (Backpacker Field Guide Series). $9.95pa. ISBN 0964667401pa.

     95 species; keys; black-and-white illustrations of leaves, flowers, and buds. Illustrated by Olivia Petrides.

513. Petrides, George A. **Trees of the Pacific Northwest: Including Oregon, Idaho, Northwest Montana, British Columbia, Yukon, and Alaska**. Williamston, MI: Explorer, 1998. 100p. (Backpacker Field Guide Series). $12.95pa. ISBN 096466741Xpa.

     Not seen; see series entry 9. Illustrated by Olivia Petrides.

514. Petrides, George A. **Trees of the Rocky Mountains: Including all Trees Growing Wild from Alaska and Yukon to Arizona and West Texas**. Williamston, MI: Explorer, 1998. 104p. (Backpacker Field Guide Series). $12.95pa. ISBN 0964667428pa.

     Not seen; see series entry 9. Illustrated by George A. Petrides and Olivia Petrides.

515. Raven, Peter H. **Native Shrubs of Southern California**. Berkeley, CA: University of California, 1966. 132p. (California Natural History Guides, No. 15). $10.95pa. ISBN 0520010507pa.

     350 species; brief key; line drawings, some color illustrations; checklist. Illustrated by Jean Colton.

516. Stukey, Maggie, and George Palmer. **Western Trees**. Helena, MT: Falcon, 1998. 144p. (A Falcon Guide). $17.95pa. ISBN 1560446234pa.

     50 species; line drawings and silhouettes; maps; emphasis on ecology. Covers northwestern U.S. Revised edition of *Western Treebook: A Field Guide for Weekend Naturalists* by George Palmer and Martha Stuckey. Illustrated by Keith Bowers.

517. Thomas, John Hunter, and Dennis R. Parnell. **Native Shrubs of the Sierra Nevada**. Berkeley, CA: University of California Press, 1974. 127p. (California Natural History Guides, No. 34). ISBN 0520027388; 0520025385pa.

     300 species; line drawings, some color illustrations.

518. Viereck, Leslie A., and Elbert L. Little. **Guide to Alaska Trees**. Washington, DC: U.S. Dept. of Agriculture, Forest Service, 1974. 98p. (Agriculture Handbook United States Department of Agriculture, No. 472).

     128 species; key; line drawings. Supersedes Agriculture Handbook No. 5, *Pocket Guide to Alaska Trees*. Based on the authors' *Alaska Trees and Shrubs*.

519. Watts, May Theilgaard, and Tom Watts. **Desert Tree Finder: A Pocket Manual for Identifying Desert Trees**. Berkeley, CA: Nature Study Guild, 1974. 59p. $3.00pa. ISBN 0912550074pa.

     60 species; key throughout; line drawings; maps; pocket size. Covers Arizona, California, and New Mexico.

520. Watts, Tom. **California Tree Finder: A Pocket Manual for Identifying California Trees**. Berkeley, CA: Nature Study Guild, 1963. 60p.

     92 species; key throughout; black-and-white illustrations; maps. Pocket sized.

521. Watts, Tom. **Pacific Coast Tree Finder: A Pocket Manual for Identifying Pacific Coast Trees**. Berkeley, CA: Nature Study Guild, 1973. 59p. $3.00pa. ISBN 0912550066pa.

     127 species; key throughout; line drawings; maps; pocket sized.

522. Watts, Tom. **Rocky Mountain Tree Finder: A Pocket Manual for Identifying Rocky Mountain Trees**. Berkeley, CA: Nature Study Guild, 1972. 59p. $3.00pa. ISBN 0912550058pa.

     79 species; key throughout; black-and-white illustrations, maps; pocket sized.

523. Whitman, Ann, ed. **Familiar Trees of North America: Western Region**. New York: Knopf, 1986. 192p. (Audubon Society Pocket Guides). $9.00pa. ISBN 0394748522pa.

     80 species; color photos of bark and leaves; silhouettes. Covers area from eastern edge of Rockies to Pacific Coast.

524.   Wilkinson, Kathleen. **Trees and Shrubs of Alberta**. Edmonton, Alta.: Lone Pine, 1990. 191p. (A Habitat Field Guide). $15.95pa. ISBN 0919433391pa.

  77 species; key to conifers; line drawings and color photos; includes information on use.

Wildflowers

This chapter covers field guides to wildflowers, herbaceous plants with conspicuous flowers growing wild. A few guides to cultivated annuals and perennials are also included if they follow the standard field guide format. Wildflower guides are second only to bird guides in terms of sheer numbers. Many are arranged in taxonomic order, which requires the user to realize that little five-petaled flowers might be found in several places. Other guides are organized by the flower color, which is easier for beginners, but may be nearly as confusing because flower color is highly variable in some species. Other arrangements are by habitat, altitude, or flowering season, all of which are also variable.

Other plants such as cacti, which are not herbaceous, or weeds, which do not have conspicuous flowers, are covered in Chapter 3, "Plants." Flowering shrubs, whether wild or cultivated, are found in Chapter 7 with other trees and shrubs. For more wildflower guides, see Chapter 2, "Flora and Fauna." Many, if not most, of the guides in Chapter 3 also include wildflowers. Most are not cross-referenced because that would result in a very large number of duplicate entries.

NORTH AMERICA

525. Case, Frederick W., Jr., and Roberta B. Case. **Trilliums**. Portland, OR: Timber, 1997. 285p. $29.95. ISBN 0881923745.

    43 species, including 6 Asian species; color photos and line drawings; maps. Species accounts include synonymy, multiple common names, description, season, distribution, habitat, varieties, and comments. Part field guide and part garden handbook.

526. Cuthbert, Mabel Jaques. **How to Know the Fall Flowers: Pictured-Keys for Determining the More Common Fall-Flowering Herbaceous Plants with Suggestions and Aids for Their Study.** Dubuque, IA: W. C. Brown, 1948. 199p. (Pictured Key Nature Series). $25.80pa. ISBN 069704811X; 0697048101pa.

     350 species; key; line drawings.

527. Hood, Susan. **National Audubon Society First Field Guide to Wildflowers.** New York: Scholastic, 1998. 160p. $17.95; $10.95 (flexi). ISBN 0590054643; 0590054864 (flexi).

     50 common species covered in detail, another 125 illustrated and described; color photos; extensive introductory material; children's guide. Includes a separate pocket-sized card illustrating the 50 main species.

528. Mohlenbrock, Robert H. **Wildflowers: A Quick Identification Guide to the Wildflowers of North America.** New York: Macmillan, 1987. 203p. (Macmillan Field Guides). ISBN 002063420Xpa.

     304 species; color photos; arranged by color.

529. Parker, Richard. **Wildflowers.** Miami, FL: Windward, 1986. 126p. $7.95pa. ISBN 0893170348pa.

     225 species; color photos; arranged by color; autumn flowers.

530. Parsons, Frances Theodora (Smith) Dana. **How to Know the Wild Flowers; A Guide to the Names, Haunts, and Habits of Our Common Wild Flowers.** Boston: Houghton Mifflin, 1989. 287p.

     585 species; line drawings and color illustrations; natural history. Illustrated by Marion Satterlee.

531. Petrie, William. **Guide to Orchids of North America.** Blaine, WA: Hancock House, 1981. 128p. $9.95pa. ISBN 0888390890pa.

     105 species; color photos and line drawings; maps. Arranged by color.

532. Reader's Digest Editors. **North American Wildlife: Wildflowers.** New York: Reader's Digest, 1998. 468p. $16.95. ISBN 0762100346pa.

     700 species; color illustrations; maps; chart listing flowers by color and flower type.

533. Ruggiero, Michael. **Spotter's Guide to Wild Flowers of North America.** 1st American ed. New York: Mayflower, 1979. 64p. ISBN 0831794240; 0831794259pa.

     160 species; color illustrations; pocket sized. Illustrated by Joyce Bee and Will Giles.

534. Venning, Frank D. **Wildflowers of North America: A Guide to Field Identification**. New York: Golden, 1984. 340p. (Golden Field Guide Series). $11.95pa. ISBN 0307470075; 0307136647pa.

   1,553 species; color illustrations; arranged by family; text and illustrations integrated. Covers North America north of Mexico. Pages crowded with illustrations; advantages include small size and complete geographical coverage. Also published in Spanish by Editorial Trillas, 1992. Illustrated by Manabu C. Saito.

535. Verhoek, Susan. **How to Know the Spring Flowers**. 2nd ed. Dubuque, IA: W. C. Brown, 1982. 244p. (Pictured Key Nature Series). $25.80pa. ISBN 0697047822pa.

   505 species; key throughout; line drawings; maps. First edition by Mabel Jaques Cuthbert.

536. Williams, John George, and Andrew E. Williams. **Field Guide to Orchids of North America: From Alaska, Greenland, and the Arctic, South to the Mexican Border**. New York: Universe, 1983. 143p. ISBN 0876634153; 0876635869pa.

   229 species; color illustrations. Illustrated by Norman Arlott.

537. Zim, Herbert Spencer, and Alexander C. Martin. **Flowers: A Guide to Familiar American Wildflowers**. New York: Golden, 1987. 159p. (Golden Guide). $5.95pa. ISBN 030764054X; 0307240541pa.

   200 species; color illustrations; arranged by color. Illustrated by Rudolf Freund.

Eastern North America

538. Adams, Kevin, and Marty Casstevens. **Wildflowers of the Southern Appalachians: How to Photograph and Identify Them**. Winston-Salem, NC: J. F. Blair, 1996. 257p. $26.95. ISBN 0895871432.

   200 species; color photos; each entry includes description, habitat, and photography tips; has about 50 pages of introductory material on wildflower photography.

539. Ajilvsgi, Geyata. **Wildflowers of Texas**. 3rd ed. Fredericksburg, TX: Shearer, 1988. 414p. $19.95; $14.95pa. ISBN 0940672154; 0940672464pa.

   378 species; color photos; arranged by color.

540. Ajilvsgi, Geyata. **Wild Flowers of the Big Thicket, East Texas, and Western Louisiana**. College Station: Texas A&M University Press, 1979. 360p. (W. L. Moody, Jr., Natural History Series, No. 4). $17.50; $14.95pa. ISBN 089096064X; 890960658pa.

   450 species described, 520 illustrated with color photos; photos arranged by habitat, descriptions by family.

541. Alderman, J. Anthony. **Wildflowers of the Blue Ridge Parkway**. Chapel Hill: University of North Carolina Press, 1997. 222p. $12.95pa. ISBN 0807846511pa.

   275 species described, 200 illustrated; color photos; arranged by color; includes 75 best viewing sites. Descriptions include general notes and use.

542. Archbald, David, Rosemary V. Fleming, and Virginia M. Kline. **Quick-Key Guide to Wildflowers: Wildflowers of Northeastern and Central United States and Adjacent Canada**. Garden City, NY: Doubleday, 1968. (unpaged). (Quick-Key Field Identification Guides, Q-K 4).

   545 species; includes a series of punch cards for "mechanized" identification as well as black–and–white line drawings and brief descriptions. Illustrated by Ellen Archbald Davis.

543. Batson, Wade T. **Wild Flowers in the Carolinas**. Columbia, SC: University of South Carolina Press, 1987. 153p. $24.95. ISBN 0872495043; 0872495051pa.

   525 species; color photos; arranged by family.

544. Bell, C. Ritchie, and Bryan J. Taylor. **Florida Wild Flowers and Roadside Plants**. Chapel Hill, NC: Laurel Hill, 1982. 308p. $19.95; $14.95pa. ISBN 0960868801; 0960868836pa.

   500 species; color photos; maps.

545. Brown, Clair A. **Wildflowers of Louisiana and Adjoining States**. Baton Rouge: Louisiana State University Press, 1972. 247p. $14.95pa. ISBN 0807102328; 0807107808pa.

   450 species; color photos.

546. Brown, Paul Martin. **Wild Orchids of the Northeastern United States: A Field and Study Guide to the Orchids Growing Wild in New England, New York, and Adjacent Pennsylvania and New York**. Ithaca, NY: Comstock, 1997. 236p. $17.95pa. ISBN 0801483417pa.

71 species and varieties; illustrated key to genera plus keys to species; color photos and black-and-white illustrations; maps; checklists; notes on where to find orchids. Descriptions include range, flowering season, habitat, degree of rarity, and general notes.

547. Chapman, William K. **Orchids of the Northeast: A Field Guide**. Syracuse, NY: Syracuse University Press, 1996. 200p. $39.95; $17.95pa. ISBN 0815626975; 0815603428pa.

    85 species or varieties; illustrated keys; color photos. Species accounts include description, etymology of scientific name, numerous common names, season, range, and comments. A fairly technical presentation. The ranges covered are the New England states and New York.

548. Chapman, William K. **Wildflowers of New York in Color**. Syracuse, NY: Syracuse University Press, 1998. 168p. $59.95; $24.95pa. ISBN 0815627467; 081560470X.

    300 species; color photos; arranged by color and type. Includes habitat and comments.

549. Core, Earl Lemley. **Spring Wild Flowers of West Virginia**. 3rd ed. Morgantown, WV: West Virginia University Press, 1981. 104p. $5.00pa. ISBN 0937058025pa.

    250 species; line drawings; arranged by family. Illustrated by William A. Lunk.

550. Courtenay, Booth, and James Hall Zimmerman. **Wildflowers and Weeds**. New York: Van Nostrand Reinhold, 1972. 144p. $12.95pa. ISBN 0671765760pa.

    600 species; "chart key"; color photos; covers Great Lakes region.

551. Dean, Blanche Evans, Amy Mason, and Joab L. Thomas. **Wildflowers of Alabama and Adjoining States**. University: University of Alabama Press, 1973. 230p. $19.95pa. ISBN 0817312005; 081730147Xpa.

    400 common species; color photos.

552. Denison, Edgar. **Missouri Wildflowers: A Field Guide to Wildflowers of Missouri and Adjacent Areas**. Rev. and expanded 4th ed. Jefferson City, MO: Missouri Dept. of Conservation, 1989. 314p. $11.50pa. ISBN 020010022pa.

    250 species illustrated by color photos with another 150 species described; arranged by color.

553. DiGregorio, Mario, and Jeff Wallner. **A Vanishing Heritage: Wildflowers of Cape Cod**. Missoula, MT: Mountain, 1989. 169p. $10.00pa. ISBN 0878422315pa.

    75 species; color photos; natural history; arranged by habitat.

554. Duncan, Wilbur Howard, and Leonard E. Foote. **Wildflowers of the Southeastern United States**. Athens: University of Georgia Press, 1975. 296p. ISBN 082030347X.

    1,100 species, half illustrated by color photos.

555. Enquist, Marshall. **Wildflowers of the Texas Hill Country**. Austin, TX: Lone Star Botanical, 1987. 275p. $14.95pa. ISBN 0961801301pa.

    427 species; color photos. Includes flowering trees and shrubs.

556. Farrar, Jon. **Field Guide to Wildflowers of Nebraska and the Great Plains**. Lincoln, NE: Nebraskaland Magazine; Nebraska Game and Parks Commission, 1990. 215p. $17.95. ISBN 096259590X.

    375 species; color photos; arranged by color.

557. Feilberg, Jon, Bent Fredskild, and Sune Holt. **Gronlands Blomster— Flowers of Greenland**. 2. opl. Ringsted, Denmark: A. Flensborgs forlag, 1984. 100p. ISBN 8799713071.

    154 species; color photos. In Danish and English.

558. Freeman, Craig Carl, and Eileen K. Schofield. **Roadside Wildflowers of the Southern Great Plains**. Lawrence, KS: University Press of Kansas, 1991. 280p. $29.95; $17.95pa. ISBN 0700604472; 0700604480pa.

    250 species; color photos; covers area centered on Kansas. Arranged by color and flowering season.

559. Good, Mary B. **Trillium: A Guide to the Common Wildflowers of Northeastern Wisconsin**. Woodruff, WI: M. B. Good, 1990. 69p. $9.95pa. ISBN 096279760Xpa.

    60 species; line drawings and color photos. Illustrated by Linda Schroeder-Golding.

560. Gupton, Oscar W., and Fred C. Swope. **Fall Wildflowers of the Blue Ridge and Great Smoky Mountains**. Charlottesville: University Press of Virginia, 1987. 208p. $14.95. ISBN 0813911230.

    100 species illustrated by color photos with another 124 species described; arranged by color of berries or seed pods; illustrations are for berries and seed pods, not flowers.

561. Gupton, Oscar W., and Fred C. Swope. **Wild Orchids of the Middle Atlantic States**. Knoxville: University of Tennessee Press, 1986. 112p. $16.95. ISBN 0870495097.

52 species; color photos; arranged by color; covers Delaware, Kentucky, Maryland, Tennessee, Virginia, and West Virginia.

562. Gupton, Oscar W., and Fred C. Swope. **Wildflowers of the Shenandoah Valley and Blue Ridge Mountains**. Charlottesville: University Press of Virginia, 1979. 208p. $14.95. ISBN 0813901840.

200 species illustrated with color photos with another 85 species described; arranged by color.

563. Gupton, Oscar W., and Fred C. Swope. **Wildflowers of Tidewater Virginia**. Charlottesville: University Press of Virginia, 1982. 208p. $14.95. ISBN 0813909228.

200 species illustrated with color photos with another 300 species described; arranged by flower color.

564. Haller, Karen S. **Walking with Wildflowers: A Field Guide to the St. Louis Area**. Columbia: University of Missouri Press, 1994. 257p. $22.50pa. ISBN 0826209505pa.

70 species found at 28 sites near St. Louis; color photos; checklist.

565. Harris, Stuart K., Jean H. Langenheim, and Frederick L. Steele. **A. M. C. Field Guide to Mountain Flowers of New England**. Boston: Appalachian Mountain Club, 1977. 147p. ISBN 0910146128pa.

149 species; color photos. Originally published as *Mountain Flowers of New England*.

566. Hemmerly, Thomas E. **Wildflowers of the Central South**. Nashville, TN: Vanderbilt University Press, 1990. 121p. $13.95pa. ISBN 0826512402pa.

150 species; color photos; arranged by family; includes information on use. Area covered is primarily central Tennessee, plus parts of Kentucky, Mississippi, and Alabama.

567. Henn, Robert L. **Wildflowers of Ohio**. Bloomington: Indiana University Press, 1998. 215p. $14.95pa. ISBN 025333695; 0253211670pa.

286 species; color photos; arranged by color and taxonomic order; descriptions include habitat, blooming period, range in Ohio, and use. Features color tabs indicating flower color. Bound so that the book is held horizontally rather than vertically.

568. Homoya, Michael A. **Orchids of Indiana**. Bloomington: Indiana University Press: Distributed by Indiana University Press, 1993. 276p. $34.95. ISBN 0253328640.

     43 species; color and black-and-white photographs; field guide/manual.

569. Hunter, Carl G. **Wildflowers of Arkansas**. 3rd ed. Little Rock, Ark.: Ozark Society Foundation, 1992. 296p. $32.95; $24.95pa. ISBN 0912456175pa; 0912456167.

     484 species illustrated with color photos with another 116 described; arranged by family. Illustrated by Bruce Cook and Kerry Cook.

570. Hutson, Robert W., Carlos C. Campbell, and Aaron J. Sharp. **Campbell, Hutson, and Sharp's Great Smoky Mountains Wildflowers**. 5th ed. Northbrook, IL: Windy Pines, 1995. 144p. $10.95pa. ISBN 0964341735pa.

     225 species; color photos; spiralbound. Revised edition of the authors' *Great Smoky Mountains Wildflowers*.

571. Janke, Robert A. **The Wildflowers of Isle Royale National Park**. rev. 2nd ed. Houghton, MI: Isle Royale Natural History Association, 1996. 96p. $5.95pa. ISBN 0935289089pa.

     101 species; black-and-white illustrations; arranged by colors of flowers. Includes table of fruits (e.g., rose hips) of common flowers. Also published by the National Park Service. Illustrated by Nadine Janke.

572. Justice, William S., and C. Ritchie Bell. **Wild Flowers of North Carolina**. Chapel Hill: University of North Carolina Press, 1968. 217p. $15.95pa. ISBN 0807841927pa.

     400 species; color photos. Reprinted 1989.

573. Kannowski, Paul Bruno. **Wildflowers of North Dakota**. Grand Forks: University of North Dakota Press, 1989. 126p. $12.95pa. ISBN 0960870032pa.

     159 species; color photos; maps; arranged by color and flowering time.

574. Kavasch, E. Barrie. **Introducing Eastern Wildflowers**. Killingworth, CT: Hancock House, 1982. 32p. (Northeast Color Series). $3.50pa. ISBN 0888390920pa.

     60 species; color photos. Pamphlet.

575. Kirkpatrick, Zoe Merriman. **Wildflowers of the Western Plains: A Field Guide**. Austin: University of Texas Press, 1992. 256p. (A Corrie Herring Hooks Book). $35.00; $16.95pa. ISBN 0292790619; 0292790627pa.

   450 common species; color photos; arranged by family; includes general notes.

576. Ladd, Douglas M., and Frank Oberle. **Tallgrass Prairie Wildflowers: A Field Guide**. Helena, MT: Falcon, 1995. 272p. $25.00pa. ISBN 1560442999pa.

   295 species including grasses; color photos; arranged by color; includes list of remaining tallgrass prairies. Includes grasses.

577. Lamoureux, Gisele, and Claude Allard, eds. **Plantes Sauvages Printanières**. Saint-Augustin, Quebec: Fleurbec, 1992. (Guide d'Identification Fleurbec). ISBN 2920174118.

   100 species; color photos; natural history; arranged by color; common names in English and French. In French.

578. Leake, Henderson, and Dorothy Leake. **Wildflowers of the Ozarks**. Little Rock, AR: The Ozark Society Foundation, 1981. 170p. $9.95pa. ISBN 0912456043pa.

   167 species; line drawings.

579. Levine, Carol. **A Guide to Wildflowers in Winter: Herbaceous Plants of Northeastern North America**. New Haven: Yale University Press, 1995. 368p. $40.00; $20.00pa. ISBN 0300062079; 0300065604pa.

   391 species covered in detail, another 191 mentioned; line drawings; illustrated keys; arranged by type (seeds, pods, some by family). Illustrated by Dick Rauh and Redenta Soprano.

580. Loughmiller, Campbell, and Lynn Loughmiller. **Texas Wildflowers: A Field Guide**. Austin: University of Texas Press, 1984. 271p. $27.95; $12.95pa. ISBN 0292780591; 0292780605pa.

   300 species; color photos.

581. Lund, Harry C. **Michigan Wildflowers in Color**. Holt, MI: Thunder Bay, 1998. 144p. $18.95pa. ISBN 0961481803pa.

   278 species; color photos; arranged by color and season. Has about 6 species per page, with photos facing descriptions.

582. Magley, Beverly. **Minnesota Wildflowers: A Children's Field Guide to the State's Most Common Flowers**. Helena, MT: Falcon, 1992. 32p. (Interpreting the Great Outdoors). $5.95pa. ISBN 1560441178pa.

66 species; color illustrations; arranged by habitat. An attractive children's guide.

583. Magley, Beverly. **North Carolina Wildflowers: A Children's Field Guide to the State's Most Common Flowers**. Helena, MT: Falcon, 1993. 32p. (Interpreting the Great Outdoors). $5.95pa. ISBN 1560441844pa.

See series entry 50.

584. Magley, Beverly. **Texas Wildflowers: A Children's Field Guide to the State's Most Common Flowers**. Billings, MT: Falcon, 1993. 32p. (Interpreting the Great Outdoors). $5.95pa. ISBN 1560441836pa.

See series entry 50.

585. Martin, Edwin M. **A Beginner's Guide to Wildflowers of the C&O Towpath**. Washington, DC: Smithsonian Institution Press, 1984. 57p. ISBN 0874746574pa.

158 species; color photos separate from text; arranged by color. Covers area from Georgetown to Cumberland, Maryland along the Chesapeake and Ohio Canal.

586. McGrath, Anne, with Joanne Treffs. **Wildflowers of the Adirondacks**. Sylvan Beach, NY: North Country, 1981. 109p. $14.95pa. ISBN 0932052274pa.

145 species; color photos separate from text; natural history; arranged by color and time of blooming.

587. Moyle, John Briggs, and Evelyn W. Moyle. **Northland Wild Flowers: A Guide for the Minnesota Region**. Minneapolis: University of Minnesota Press, 1977. 236p. $18.95pa. ISBN 0816608067; 0816613559pa.

300 species illustrated by color photos, another 100 species described; arranged by color.

588. Newcomb, Lawrence. **Newcomb's Wildflower Guide: An Ingenious New Key System for Quick, Positive Field Identification of the Wildflowers, Flowering Shrubs and Vines of Northeastern and North Central North America**. Boston: Little, Brown, 1977. 490p. $18.95pa. ISBN 0316604410; 0316604429pa.

1,375 species; key; color and black-and-white illustrations. Illustrated by Gordon Morrison.

589. Niehaus, Theodore F. **A Field Guide to Southwestern and Texas Wildflowers**. Boston: Houghton Mifflin, 1984. 449p. (Peterson Field Guide Series, No. 31). $17.95pa; $20.95. ISBN 0395366402pa; 0395328764.

     1,505 species; key to families; color illustrations and line drawings; arranged by color and form. Illustrated by Charles L. Ripper and Virginia Savage.

590. Niering, William A., and Nancy C. Olmstead. **The Audubon Society Field Guide to North American Wildflowers, Eastern Region**. New York: Knopf, 1979. 863p. (Audubon Society Field Guide Series). $19.00 (flexi). ISBN 0394504321 (flexi).

     600 species covered in detail, another 400 briefly described; color photos; arranged by flower color.

591. Owensby, Clenton E. **Kansas Prairie Wildflowers**. Ames: Iowa State University Press, 1980. 124p. $19.95pa. ISBN 0813808502; 0813811600pa.

     200 common species; keys; color photos; maps; arranged by color.

592. Peterson, Roger Tory. **Peterson First Guide to Wildflowers of Northeastern and North-Central North America**. Boston: Houghton Mifflin, 1986. 126p. $5.95pa. ISBN 0395906679pa.

     188 species; color illustrations; arranged by color. Illustrated by Roger Tory Peterson.

593. Peterson, Roger Tory, and Margaret McKenny. **A Field Guide to Wildflowers of Northeastern and North-Central North America: A Visual Approach Arranged by Color, Form, and Detail**. Boston: Houghton Mifflin, 1968. 420p. (Peterson Field Guide Series, No. 17). $18.00pa. ISBN 039508086X; 0395911729pa.

     1,293 species; line drawings and color illustrations; arranged by color and type of flower. Illustrated by Roger Tory Peterson.

594. Porcher, Richard D. **Wildflowers of the Carolina Lowcountry and Lower Pee Dee**. Columbia, SC: University of South Carolina Press, 1995. 344p. $24.95pa. ISBN 1570030944; 1570030278pa.

     437 species; color photos separate from text; natural history; arranged by habitat. Also includes information on 48 natural areas and edible and poisonous plants.

595. Roberts, June Carver. **Born in the Spring: A Collection of Spring Wildflowers**. Athens: Ohio University Press, 1976. 159p. $21.95pa. ISBN 0821401955; 0821402269pa.

     200 species; color illustrations; arranged by flowering date.

596. Roland, A. E. **Spring Wildflowers**. Halifax, N.S.: Nimbus, 1993. 138p. (Field Guide Series). ISBN 1551090503.

    250 species; line drawings; includes flowering trees, shrubs, and forbs; arranged by season and type of plant. Covers plants of Nova Scotia.

597. Rose, Francis L. **Wildflowers of the Llano Estacado**. Dallas, TX: Taylor, 1986. 100p. $12.95. ISBN 0961710209.

    117 species illustrated with color photos; key to families; extensive background. Texas flowers.

598. Runkel, Sylvan T., and Alvin F. Bull. **Wildflowers of Illinois Woodlands**. 1st Iowa State University Press ed. Ames: Iowa State University Press, 1994. 257p. $24.95pa. ISBN 0813819903pa.

    125 species; color photos; natural history and use; arranged by flowering time. Includes multiple common names and etymology of Latin names.

599. Runkel, Sylvan T., and Alvin F. Bull. **Wildflowers of Indiana Woodlands**. 1st Iowa State University Press ed. Ames: Iowa State University Press, 1994. 257p. $24.95pa. ISBN 0813819695pa.

    125 species; color photos; natural history and use; arranged by flowering time. Includes multiple common names and etymology of Latin names.

600. Runkel, Sylvan T., and Alvin F. Bull. **Wildflowers of Iowa Woodlands**. 1st Iowa State University Press ed. Ames: Iowa State University Press, 1987. 257p. $24.95pa. ISBN 0813819296pa.

    125 species; color photos; natural history and use; arranged by flowering time. Includes multiple common names and etymology of Latin names.

601. Runkel, Sylvan T., and Dean M. Roosa. **Wildflowers of the Tallgrass Prairie: The Upper Midwest**. Ames: Iowa State University Press, 1989. 279p. $24.95pa. ISBN 0813819792pa.

    130 species; color photos; natural history and use.

602. Schinkel, Dick. **Favorite Wildflowers of the Great Lakes and the Northeastern U.S.** Lansing, MI: Thunder Bay, 1994. $16.95pa. ISBN 1882376048pa.

    100 species; color illustrations; includes information on which animals eat the plants, plus cultivation and natural history. Illustrated by David Mohrardt.

603. Seymour, Randy. **Wildflowers of Mammoth Cave National Park**. Lexington, KT: University Press of Kentucky, 1997. 254p. $17.95pa. ISBN 0813108985pa.

    400 species of trail- and roadside plants; color photos; arranged by flowering period and color. Includes description, flowering period, frequency within the park, and folklore and use of each plant. Appendixes show where within the park each plant is found, and how many species are found along the major trails.

604. Simonds, Roberta L., and Henrietta H. Tweedie. **Wildflowers of the Great Lakes Region**. 2nd ed. Champaign, IL: Stipes, 1997. 146p. ISBN 0875637213pa.

    300 species; line drawings; arranged by season and habitat. Illustrated by the authors.

605. Smith, Arlo I. **A Guide to Wildflowers of the Mid-South: West Tennessee into Central Arkansas and South Through Alabama and into East Texas**. Memphis: Memphis State University Press, 1979. 281p. $19.95. ISBN 0878700765.

    475 species; keys; color photos.

606. Smith, Richard M. **Wildflowers of the Southern Mountains**. Knoxville: University of Tennessee Press, 1998. 262p. $24.95pa. ISBN 0870499920pa.

    1,200 species; visual key to genus or family; color photos in separate plates; about half the species are illustrated. Covers the southern Appalachian Mountain region.

607. Smith, Welby R. **Orchids of Minnesota**. Minneapolis: University of Minnesota Press, 1993. 172p. $24.95. ISBN 0816623090.

    43 species; keys; black-and-white illustrations and color photos; maps of both Minnesota and North America; appendix with flowering dates. Extensive descriptions, including abundance, habitat, similar species, and comments. Illustrated by Vera M. Wong.

608. Spellenberg, Richard. **Familiar Flowers of North America: Eastern Region**. New York: Knopf, 1986. 192p. (Audubon Society Pocket Guides). $9.00pa. ISBN 0394748433pa.

    80 species; color photos; arranged by color.

609. Strauch, Joseph G. **Wildflowers of the Berkshire and Taconic Hills**. Stockbridge, MA: Berkshire House, 1995. 130p. (Berkshire Outdoors). $12.95. ISBN 093639966X.

167 species; color photos separate from text; arranged by color and season of blooming; natural history. Covers parts of western Massachusetts. Illustrated by the author.

610. Taylor, Walter Kingsley. **Florida Wildflowers in Their Natural Communities**. Gainesville: University Press of Florida, 1999. 400p. $24.95pa. ISBN 0813016169pa.

450 species; color photos; descriptions include flowering time, habitat, range, and comments. Arranged by habitat.

611. Taylor, Walter Kingsley. **The Guide to Florida Wildflowers**. Dallas, TX: Taylor, 1992. 320p. $26.95. ISBN 0878337474.

574 species; color photos; arranged by color.

612. Timme, S. Lee. **Wildflowers of Mississippi**. Jackson: University Press of Mississippi, 1989. 278p. (Natural History Series). $30.00; $22.95pa. ISBN 0878053956; 0878054847pa.

521 species described, 100 others mentioned; color photos; by family.

613. Titford, Bill, and June Titford. **A Travellers' Guide to Wild Flowers of Newfoundland, Canada**. St. John's, Nfld.: Flora Frames, 1995. 106p. ISBN 0969945906.

540 species; color photos; includes flower-finding sites in Newfoundland and short poems.

614. Tveten, John L., and Gloria Tveten. **Wild Flowers of Houston and Southeast Texas**. Austin: Texas University Press, 1997. 320p. $21.95pa. ISBN 0292781512pa.

200 species; color photos; arranged by color; includes information on use and lore; tips on where to find flowers.

615. Vance, F. R., J. R. Jowsey, and J. S. McLean. **Wildflowers of the Northern Great Plains**. 2nd, rev. ed. Minneapolis: University of Minnesota Press, 1984. 336p. $29.95; $15.95pa. ISBN 0816613508; 0816613516pa.

400 species; key; color photos and black-and-white drawings; covers Saskatchewan, Manitoba, Alberta, the Northwest Territories, Montana, the Dakotas, and Minnesota. Revised edition of *Wildflowers Across the Prairies*.

616. Wallner, Jeff, and Mario J. DiGregorio. **New England's Mountain Flowers: A High Country Heritage**. Missoula, MT: Mountain, 1997. 224p. $17.95pa. ISBN 0878423370pa.

85 species; color photos; arranged by habitat type; includes natural history and plant lore. Covers Connecticut, Massachusetts, and New York.

617.   Watts, May Theilgaard. **Flower Finder: A Key to Spring Wild Flowers and Flower Families East of the Rockies and North of the Smokies, Exclusive of Trees and Shrubs**. Berkeley, CA: Nature Study Guild, 1955. 60p. $3.00. ISBN 0912550007pa.

155 species; key throughout; line drawings; pocket size.

618.   Woodward, Carol H., and Harold William Rickett. **Common Wild Flowers of the Northeastern United States**. Woodbury, NY: Barron's, 1979. 318p. (New York Botanical Garden's Field Guide). ISBN 0812009371.

1,000 species described, 550 illustrated with color photos; arranged by family; includes flowers found within 50 miles of Manhattan; no trees, shrubs, or vines are included.

Western North America

619.   Arnberger, Leslie P. **Flowers of the Southwest Mountains**. Globe, AZ: Southwest Parks and Monuments Association, 1982. 139p. $9.95pa. ISBN 0911408614pa.

150 species; line drawings and color photos; arranged by color. Illustrated by Jeanne R. Janish.

620.   Beaudoin, Viola Kneeland. **The Beaudoin Easy Method of Identifying Wildflowers: Over 475 Mountain Wildflowers**. Aurora, CO: Evergreen, 1983. 234p. $8.95pa. ISBN 0961196009pa.

475 species; line drawings, some color photos; arranged by color and shape; covers the Rocky Mountains. Illustrated by Genevieve Berry and Elizabeth Houser.

621.   Bernard, Nelson T. **Wildflowers along Forests and Mesa Trails**. Albuquerque: University of New Mexico Press, 1984. 177p. ISBN 0826307302.

80 species; line drawings; natural history; covers area around Albuquerque and the New Mexico highlands. Arranged by color. Illustrated by Dan Godfrey.

622.   Blackwell, Laird R. **Wildflowers of the Tahoe Sierra: From Forest Deep to Mountain Peak**. Redmond, WA: Lone Pine, 1997. 144p. $9.95pa. ISBN 1551050854pa.

100 common species; color photos; arranged by habitat; descriptions include similar species. Pocket sized.

623.   Bowers, Janice Emily. **100 Roadside Wildflowers of Southwest Woodlands**. Tucson, AZ: Southwest Parks and Monuments Association, 1987. 64p. $4.95pa. ISBN 0911408738pa.
Color photos; arranged by color.

624.   Bowers, Janice Emily. **100 Desert Wildflowers of the Southwest**. Tucson, AZ: Southwest Parks and Monuments Association, 1989. 1 v. (unpaged). $4.95pa. ISBN 091140872Xpa.
Color photos; natural history; pamphlet.

625.   Buchanan, Hayle. **Wildflowers of Southwestern Utah: A Field Guide to Bryce Canyon, Cedar Breaks, and Surrounding Plant Communities**. Rev. ed. Bryce Canyon, Utah: Bryce Canyon Natural History Association; Distributed by Falcon, 1992. 119p. $6.95pa. ISBN 1560440740pa.
128 species; color photos and line drawings; includes interesting tidbits. Arranged by habitat.

626.   Burbridge, Joan. **Wildflowers of the Southern Interior of British Columbia and Adjacent Parts of Washington, Idaho, and Montana**. Vancouver, B.C.: University of British Columbia Press, 1989. 398p. $29.95; $19.95pa. ISBN 0774803258pa; 0774803207.
335 species; color photos; arranged by color, one species per page.

627.   Bush, C. Dana. **The Compact Guide to Wildflowers of the Rockies**. Edmonton, Alta.: Lone Pine, 1990. 143p. $7.95pa. ISBN 091943357Xpa.
100 species; color illustrations; for beginners.

628.   Coffeen, Mary. **Central Coast Wildflowers: Monterey, San Luis Obispo, and Santa Barbara Counties of California**. San Luis Obispo, CA: EZ Nature, 1993. 160p. $12.95pa. ISBN 0945092210pa.
150 species; line drawings, color photos separate from text; arranged by habitat; natural history.

629.   Coleman, Ronald A. **The Wild Orchids of California**. Ithaca, NY: Comstock, 1995. 129p. $45.00. ISBN 0801430127; 0801482070pa.
321 species; color photos; maps; extensive natural history.

630. Craighead, John J., Frank C. Craighead, Jr., and Ray J. Davis. **A Field Guide to Rocky Mountain Wildflowers from Northern Arizona and New Mexico to British Columbia**. Boston: Houghton Mifflin, 1991. 277p. (Peterson Field Guide Series, No. 14). $16.95pa. ISBN 0395936136pa.

     590 species; key; color photos and black-and-white drawings; descriptions include flowering season (with notes on events happening at the same time), where found (including national parks), and general notes (including use). Arranged by family. Illustrated by Grant O. Hagen and Eduardo Salgado.

631. Dale, Nancy. **Flowering Plants: The Santa Monica Mountains, Coastal and Chaparral Regions of Southern California**. Santa Barbara, CA: Capra, 1986. 239p. ISBN 0884962393pa.

     260 species; color photos; indexed by color; includes information on uses of plants and suggested botanizing trips.

632. Dodge, Natt Noyes. **Flowers of the Southwest Deserts**. Tucson, AZ: Southwest Parks and Monuments Association, 1985. 136p. $9.95pa. ISBN 0911408657pa.

     120 species; line drawings and color photos; arranged by color on colored pages. Illustrated by Jeanne R. Janish.

633. Dorward, Doreen Marsh, and Sally Randall Swanson. **Along Mountain Trails (and in Boggy Meadows): A Guide to Northern Rocky Mountain Wildflowers**. Ketchum, ID: Boggy Meadows, 1993. 104p. ISBN 1560442344pa.

     90 species; color photos and line drawings; arranged by color; includes information on use and habitat.

634. Duft, Joseph F., and Robert K. Moseley. **Alpine Wildflowers of the Rocky Mountains**. Missoula: Mountain, 1989. 200p. $14.00pa. ISBN 0878422382pa.

     300 species; color photos separate from text; arranged by color. Includes a few ferns and trees.

635. Ferris, Roxana S. **Death Valley Wildflowers**. Rev. ed. Death Valley, CA: Death Valley Natural History Association, 1983. 150p. $7.95pa. ISBN 1878900072pa.

     140 species; line drawings; natural history.

636. Ferris, Roxana S. **Flowers of Point Reyes National Seashore**. Berkeley, CA: University of California Press, 1970. 119p. $10.95pa. ISBN 0520016947pa.

200 species; black-and-white drawings; indexed by flower color. Illustrated by Jeanne R. Janish.

637. Foxx, Teralene S. **Flowers of the Southwestern Forests and Woodlands**. Los Alamos, NM: Los Alamos Historical Society, 1984. 210p. ISBN 0941232051pa.

    450 species; keys; line drawings. Covers southwest mountains from 6,000–10,000 feet. Illustrated by Dorothy Hoard.

638. Griffiths, Anthony J. F., and Fred R. Ganders. **Wildflower Genetics: A Field Guide for British Columbia and the Pacific Northwest**. Vancouver, B.C.: Flight, 1983. 215p. ISBN 091984300X.

    Not your usual field guide; covers genetic variations such as unusual colors in a number of plants of the Pacific Northwest; black-and-white illustrations.

639. Guennel, G. K. **Guide to Colorado Wildflowers**. Englewood, CO: Westcliffe, 1995. 2 v. $24.95pa (v. 1); $24.95pa (v. 2). ISBN 1565791185pa (v. 1); 1565791193pa (v. 2).

    Volume 1, Plains and Foothills; volume 2, Mountains. 300 species in each volume; color photos and illustrations; arranged by color. Includes some trees.

640. Horn, Elizabeth L. **Coastal Wildflowers of the Pacific Northwest: Wildflowers and Flowering Shrubs from British Columbia to Northern California**. Missoula, MT: Mountain, 1993. 179p. ISBN 0878422919pa.

    164 species; color photos; arranged by habitat.

641. Jensen, Earl R. **Flowers of Wyoming's Big Horn Mountains and Big Horn Basin**. Greybull, WY: E. R. Jensen, 1987. 251p. $20.00pa.

    Comprehensive; key throughout; line drawings; spiralbound. Includes trees.

642. Jolley, Russ. **Wildflowers of the Columbia Gorge: A Comprehensive Guide**. Portland, OR: Oregon Historical Society Press, 1988. 344p. (Jack Murdock Series). $24.95pa. ISBN 0875951880pa.

    744 species; color photos; blooming dates; fold-out map of gorge.

643. Jones, Charlotte Foltz. **Colorado Wildflowers: A Beginner's Field Guide to the State's Most Common Flowers**. Billings, MO: Falcon, 1994. 32p. (Interpreting the Great Outdoors). $5.95pa. ISBN 1560442662pa.

66 species; color illustrations; natural history; arranged by habitat. Oversized children's guide.

644.   Keator, Glenn. **Sierra Flower Finder: A Guide to Sierra Nevada Wildflowers**. Berkeley, CA: Nature Study Guild, 1980. $3.75pa. ISBN 0912550090pa.

45 species; key throughout; black-and-white illustration.

645.   Kinucan, Edith S., and Penney R. Brons. **Wild Wildflowers of the West**. 4th ed., rev. Clark Fork, ID: Kinucan and Brons, 1991. 133p. $9.95pa. ISBN 0961544414pa.

120 species; color photos; arranged by color; includes information on use.

646.   Larrison, Earl Junior. **Washington Wildflowers, Including 1134 Species of Wildflowers Most Commonly Found in the State of Washington and Adjacent Areas of Oregon, Idaho, and British Columbia**. Seattle, WA: Seattle Audubon Society, 1974. 376p. (Trailside Series). ISBN 0914516027pa.

Keys; color and black-and-white photos.

647.   Lyons, C. P. **Wildflowers of Washington**. Edmonton, Alta.: Lone Pine, 1997. 192p. $15.95pa. ISBN 1551050927pa.

500 species; color photos; arranged by color of flower, flower type, and size. Descriptions include range and notes, which may include mention of the species by early explorers or use by local Native Americans.

648.   Magley, Beverly. **Arizona Wildflowers: A Children's Field Guide to the State's Most Common Flowers**. Helena, MT: Falcon, 1991. 32p. (Interpreting the Great Outdoors). $5.95pa. ISBN 1560440961.

66 species; color illustrations; arranged by habitat. Includes multiple common names. An attractive children's guide.

649.   Magley, Beverly. **California Wildflowers: A Children's Field Guide to the State's Most Common Flowers**. Billings, MT: Falcon, 1989. 32p. (Interpreting the Great Outdoors). $5.95pa. ISBN 0937959588pa.

58 species; color illustrations; arranged by habitat. An attractive children's guide.

650.   Magley, Beverly. **Montana Wildflowers: A Children's Field Guide to the State's Most Common Flowers**. Helena, MT: Falcon, 1992. 32p. (Interpreting the Great Outdoors). $5.95pa. ISBN 1560441186pa.

68 species; color illustrations; arranged by habitat. An attractive children's guide.

651. Magley, Beverly. **Oregon Wildflowers: A Children's Field Guide to the State's Most Common Flowers**. Billings, MT: Falcon, 1992. 32p. (Interpreting the Great Outdoors). $5.95pa. ISBN 156044035Xpa.

Not seen; see series entry 50.

652. Manning, Harvey. **Mountain Flowers**. Seattle, WA: Mountaineers, 1979. 96p. $7.95pa. ISBN 0916890929pa.

84 of the most common flowers of the Cascades and Olympic ranges; color photos; 150 other species mentioned. Arranged by color and habitat. Pocket-sized spiralbound volume. Cover title: *Mountain Flowers of the Cascades and Olympics*.

653. Martin, William C., and Charles R. Hutchins. **Fall Wildflowers of New Mexico**. Albuquerque: University of New Mexico Press, 1988. 294p. (New Mexico Natural History Series). ISBN 082631080X; 0826310818pa.

348 species; keys; line drawings, some color photos. Illustrated by Robert D. Ivey.

654. Martin, William C., and Charles R. Hutchins. **Spring Wildflowers of New Mexico**. Albuquerque: University of New Mexico Press, 1984. 257p. (New Mexico Natural History Series). ISBN 0826307426; 0826307434pa.

366 species; keys; black-and-white drawings, some color photos.

655. Martin, William C., and Charles R. Hutchins. **Summer Wildflowers of New Mexico**. Albuquerque: University of New Mexico Press, 1986. 318p. (New Mexico Natural History Series). ISBN 0826308597pa; 0826308600.

390 species; key; black-and-white drawings, some color photos. Illustrated by Robert D. Ivey.

656. Miller, Millie. **Kinnikinnick: The Mountain Flower Book**. Boulder, CO: Johnson, 1980. 39 leaves. (Millie and Cyndi's Pocket Nature Guides). $5.95pa. ISBN 0933472099pa.

120 species; color illustrations; pocket size.

657. Miller, Millie. **Saguaro: The Desert Flower Book**. Boulder, CO: Johnson, 1982. 1 v. (unpaged). (Millie and Cyndi's Pocket Nature Guides). $5.95pa. ISBN 0933472692pa.

120 species; color illustrations; pocket size.

658. Munz, Philip Alexander. **California Desert Wildflowers**. Berkeley, CA: University of California Press, 1962. 122p. $11.95pa. ISBN 052000897; 0520008995pa.

　　300 species; color photos and black-and-white drawings; arranged by color.

659. Munz, Philip Alexander. **California Mountain Wildflowers**. Berkeley, CA: University of California Press, 1963. 122p. $11.95pa. ISBN 0520009010pa.

　　300 species; color photos and black-and-white drawings; arranged by color.

660. Munz, Philip Alexander. **California Spring Wildflowers, From the Base of the Sierra Nevada and Southern Mountains to the Sea**. Berkeley, CA: University of California Press, 1961. 122p. $12.95pa. ISBN 0520008960pa.

　　300 species; color photos and black-and-white drawings; arranged by color.

661. Munz, Philip Alexander. **Shore Wildflowers of California, Oregon, and Washington**. Berkeley, CA: University of California Press, 1964. 122p. $11.95pa. ISBN 0520009029; 0520009037pa.

　　300 species; color photos and black-and-white drawings; arranged by color.

662. Niehaus, Theodore F. **A Field Guide to Pacific States Wildflowers: Field Marks of Species Found in Washington, Oregon, California, and Adjacent Areas: A Visual Approach Arranged by Color, Form, and Detail**. Boston: Houghton Mifflin, 1976. 432p. (Peterson Field Guide Series, No. 22). $18.00pa. ISBN 0395216249; 0395910951pa.

　　1,492 species; key to families; color illustrations and line drawings; arranged by color and form. Illustrated by Charles L. Ripper.

663. Phillips, Arthur Morton, III. **Grand Canyon Wildflowers**. Rev. ed. Grand Canyon National Park: Grand Canyon Natural History Association, 1990. 145p. $16.95. ISBN 0938216015.

　　130 species; color photos; arranged by color; oversized; natural history. Illustrated by John Richardson.

664. Porsild, A. E. **Rocky Mountain Wild Flowers**. Ottawa: National Museum of Natural Sciences and Parks Canada, Dept. of Indian and Northern Affairs, 1974. 454p. (National Museum of Natural

Sciences. Botany Division. Natural History Series, No. 2). $9.95pa. ISBN 0226564959pa.

430 species described, 250 illustrated; color illustrations; for Jasper, Banff, and Waterton Lakes National Parks in Canada.

665. Pratt, Verna E. **Field Guide to Alaskan Wildflowers**. Anchorage, AK: Alaskakrafts, 1989. 136p. $13.95pa. ISBN 0962319201pa.

200 species; color photos; arranged by color; includes good wildflower sites.

666. Pratt, Verna E. **Wildflowers Along the Alaska Highway from Dawson Creek, BC to Delta Junction and on to Fairbanks, AK**. Anchorage, AK: Alaskakrafts, 1991. 230p. $22.95pa. ISBN 096231921Xpa.

450 species; color photos; arranged by color; includes lists of species found along particular stretches of the road.

667. Pratt, Verna E., and Frank G. Pratt. **Wildflowers of Denali National Park**. Anchorage, AK: Alaskakrafts, 1993. 166p. $19.95pa. ISBN 0962319228pa.

250 species; color photos; includes mosses, fungi, lichens; arranged by color; includes checklists of species found in particular areas of the park.

668. Ross, Robert A., and Henrietta L. Chambers. **Wildflowers of the Western Cascades**. Portland, OR: Timber, 1988. 140p. $19.95pa. ISBN 0881920789pa.

475 species; color photos and line drawings. Emphasizes plants found in the Iron Mountain area in Oregon, but useful throughout the region.

669. Royer, France, and Richard Dickinson. **Wild Flowers of Edmonton and Central Alberta**. Edmonton, Alta.: University of Alberta Press, 1996. 119p. $14.95pa. ISBN 0888642822pa.

100 species; key; color photos; arranged by family. Includes description, flowering period, habitat, similar species, and notes on use and edibility.

670. Royer, France, and Richard Dickinson. **Wildflowers of Calgary and Southern Alberta**. Edmonton, Alta.: University of Alberta Press, 1996. 119p. ISBN 0888642830pa.

103 species described and illustrated, another 57 mentioned; color photos; arranged by family. Descriptions include where, when, and what to look for, plus general notes.

671. Schreier, Carl. **A Field Guide to Wildflowers of the Rocky Mountains**. Moose, WY: Homestead, 1996. 224p. $18.95pa. ISBN 0943972132pa.

    300 species; key to families; color photos and black-and-white illustrations; arranged by family. Descriptions include range and use. Covers the area from Alaska to the Mexican border.

672. Scotter, George Wilby, and Halle Flygare. **Wildflowers of the Canadian Rockies**. Edmonton, Alta: Hurtig, 1986. 170p. ISBN 088830286Xpa; 0888302851.

    200 species; color photos; arranged by color.

673. Sharsmith, Helen K. **Spring Wildflowers of the San Francisco Bay Region**. Berkeley, CA: University of California Press, 1965. 192p. (California Natural History Guides, No. 11). $10.95pa. ISBN 0520030982; 0520011686pa.

    300 species; key; line drawings and color photos; natural history.

674. Shaw, Richard J. **Utah Wildflowers: A Field Guide to Northern and Central Mountains and Valleys**. Logan, UT: Utah State University Press, 1995. 218p. $12.95pa. ISBN 0874211700pa.

    102 species; color photos; arranged by color.

675. Shaw, Richard J. **Wildflowers of Grand Teton and Yellowstone National Parks**. Rev. ed. Salt Lake City: Wheelwright, 1992. 64p. ISBN 0937512052.

    108 species; color photos; arranged by color; pamphlet. Emphasis is on flowers found along roads and hiking trails.

676. Shaw, Richard J. **Wildflowers of the Wasatch and Uinta Mountains**. Rev. ed. Salt Lake City: Wheelwright, 1983. 61p. ISBN 0937512044.

    92 species; color photos; arranged by color; pamphlet. Emphasis is on flowers found along roads and hiking trails.

677. Showers, Mary Ann, and David W. Showers. **A Field Guide to the Flowers of Lassen Volcanic National Park**. Mineral, CA: Loomis Museum Association, 1981. 112p.

    119 species; line drawings; arranged by habitat and color. Illustrated by Mary Ann Showers.

678. Spellenberg, Richard. **Familiar Flowers of North America: Western Region**. New York: Knopf, 1986. 192p. (Audubon Society Pocket Guides). $9.00pa. ISBN 0394748441pa.

    80 species; color photos and line drawings.

679. Spellenberg, Richard. **The Audubon Society Field Guide to North American Wildflowers, Western Region**. New York: Knopf, 1979. 862p. (Audubon Society Field Guide Series). $19.00 (flexi). ISBN 0394504313 (flexi).

     600 species described in detail, another 400 mentioned briefly; color photos.

680. Stark, Milt. **A Flower-Watcher's Guide to Spring-Blooming Wildflowers of the Antelope Valley**. Lancaster, CA: Flower-watcher, 1991. 159p.

     200 species; color photos separate from text; arranged by color.

681. Stewart, Charles. **Wildflowers of the Olympics and Cascades**. Rev. ed. Port Angeles, WA: Nature Education Enterprises, 1994. 1 v. (unpaged). $11.95pa. ISBN 0962110426pa.

     224 species illustrated, 40 mentioned; color photos.

682. Stewart, Jon Mark. **Colorado Desert Wildflowers: A Guide to Flowering Plants of the Low Desert, Including the Coachella Valley, Anza-Borrego Desert, and Portions of Joshua Tree National Monument**. Palm Desert, CA: J. Stewart Photography, 1993. 120p. $12.95pa. ISBN 0963490907pa.

     100 common species; color photos; brief descriptions often include use; arranged by color, cacti separate.

683. Stewart, Jon Mark. **Mojave Desert Wildflowers: A Guide to High Desert Wildflowers of California, Nevada, and Arizona**. Albuquerque, NM: J. Stewart Photography, 1998. 210p. $14.95pa. ISBN 0963490915pa.

     195 species; color photos; arranged by color. Includes separate chapter on cacti.

684. Stocking, Stephen K., and Jack A. Rockwell. **Wildflowers of Sequoia and Kings Canyon National Parks**. Rev. 1989. Three Rivers, CA: Sequoia Natural History Association, 1989. 48p.

     175 species; color photos; arranged by color. Descriptions include elevation where found and use.

685. Strickler, Dee. **Alpine Wildflowers: Showy Wildflowers of the Alpine and Subalpine Areas of the Northern Rocky Mountain States**. Columbia Falls, MT: Flower, 1990. 112p. $9.95pa. ISBN 1560440112pa.

     200 species; color photos; descriptions include habitat, range, and comments. In taxonomic order.

686. Strickler, Dee. **Forest Wildflowers: Showy Wildflowers of the Woods, Mountains, and Forests of the Northern Rocky Mountain States**. Columbia Falls, MT: Flower, 1988. 96p. $9.95pa. ISBN 0937959367pa.

150 species; color photos. In taxonomic order.

687. Strickler, Dee. **Northwest Penstemons: 80 Species of Penstemon Native to the Pacific Northwest**. Columbia Falls, MT: Flower, 1997. 191p. $29.95pa. ISBN 1560445726pa.

80 species; key; color photos and line drawings; maps. Detailed descriptions use technical language. Illustrated by Anne Morley.

688. Strickler, Dee. **Prairie Wildflowers: Showy Wildflowers of the Plains, Valleys, and Foothills in the Northern Rocky Mountain States**. Columbia Falls, MT: Flower, 1986. 80p. $9.95pa. ISBN 093431899pa.

125 species; color photos. In taxonomic order.

689. Strickler, Dee. **Wayside Wildflowers of the Pacific Northwest**. Columbia Falls, MT: Flower, 1993. 272p. $19.95pa. ISBN 1560441852pa.

400 species; color photos, line drawings; includes description, range, and habitat; one to two species per page.

690. Taylor, Ronald J. **Desert Wildflowers of North America**. Missoula, MT: Mountain, 1998. 249p. $25.00pa. ISBN 0878423761pa.

500 species; key to families; color photos; maps.

691. Taylor, Ronald J. **Rocky Mountain Wildflowers**. 2nd ed. Seattle, WA: Mountaineers, 1986. 94p. $7.95. ISBN 0898861314pa.

145 common species; key; color photos; arranged by color and flower shape, then habitat. Pocket sized.

692. Taylor, Ronald J. **Sagebrush Country: A Wildflower Sanctuary**. Rev. ed. Missoula, MT: Mountain, 1992. 211p. $14.00pa. ISBN 0878422803pa.

300 species; color photos.

693. Trelawny, John G. S. **Wildflowers of the Yukon, Alaska, and Northwestern Canada**. Rev. ed. Victoria, B.C., Canada: Sono Nis, 1988. 214p. ISBN 0919203957pa.

300 species; color photos; by family, indexed by color.

694. Underhill, J. E. (Ted). **Alpine Wildflowers**. Blaine, WA: Hancock House, 1986. 61p. $6.95pa. ISBN 0888399758pa.

    73 species; color photos and line drawings; arranged by color. Very attractive. Covers the Pacific Northwest.

695. Underhill, J. E. (Ted). **Coastal Lowland Wildflowers**. Blaine, WA: Hancock House, 1986. 64p. ISBN 0888399731.

    73 species; color photos and line drawings; arranged by color. Very attractive. Covers the Pacific Northwest.

696. Underhill, J. E. (Ted). **Roadside Wildflowers of the Northwest**. Vancouver, B.C.: Hancock House, 1981. 48p. $5.95pa. ISBN 0888391080pa.

    73 species; color photos and line drawings; arranged by color. Very attractive. Covers the Pacific Northwest.

697. Underhill, J. E. (Ted). **Sagebrush Wildflowers**. Blaine, WA: Hancock House, 1986. 64p. $6.95pa. ISBN 0888391714pa.

    73 species; color photos and line drawings; arranged by color. Very attractive. Covers the Pacific Northwest.

698. Underhill, J. E. (Ted). **Upland Field and Forest Wildflowers**. Blaine, WA: Hancock House, 1986. 64p. $4.95pa. ISBN 0888391749pa.

    73 species; color photos and line drawings; arranged by color. Very attractive. Covers the Pacific Northwest.

699. Watts, Phoebe. **Redwood Region Flower Finder**. Berkeley, CA: Nature Study Guild, 1979. $3.00pa. ISBN 0912550082pa.

    160 species; key throughout; black-and-white illustrations; covers coastal fog belt of central California to southern Oregon. Pocket sized.

700. Welsh, Stanley L. **Wildflowers of Zion National Park**. Springdale, UT: Zion Natural History Association, 1990. 136p. $9.95pa. ISBN 0915630273pa.

    120 species; color photos; natural history.

701. White, Helen A., ed. **The Alaska-Yukon Wild Flowers Guide**. Anchorage: Alaska Northwest, 1974. 218p. $19.95pa. ISBN 0882400320pa.

    160 species; color photos and line drawings; arranged by family. Illustrated by Virginia Howie.

702.   Wilkinson, Kathleen. **Wildflowers of Alberta**. Edmonton, Alta.: University of Alberta Press, 1997. 400p. ISBN 0888642989pa.

    250 species; color photos; includes information on use.

703.   Wilson, Lynn, and Jim Wilson. **Wildflowers of Yosemite**. El Portal, CA: Sierra, 1992. 143p. $9.95pa. ISBN 0939365022pa.

    224 species; color photos; arranged by color. Initial chapters describe the regions of the park and the flowers found in those regions, followed by color plates (four flowers per page), then brief species descriptions. Originally published by Sunrise Productions, 1987. Illustrated by Jeff Nicholas.

704.   Wingate, Janet L. **Rocky Mountain Flower Finder: A Guide to Wildflowers Found Below Tree Line in the Rocky Mountains**. Berkeley, CA: Nature Study Guild, 1990. 119p. $4.75pa. ISBN 0912550201pa.

    275 species; key throughout; black-and-white illustrations. Pocket sized. Illustrated by the author.

705.   Wingate, Janet L., and Loraine Yeatts. **Alpine Flower Finder: The Key to Wildflowers Found Above Treeline in the Rocky Mountains**. Boulder, CO: Roberts Rinehart, 1995. 130p. ISBN 1570980268pa.

    350 species; key; black-and-white illustrations. Illustrated by Janet L. Wingate.

706.   Young, Dorothy King. **Redwood Empire Wildflowers**. 4th ed. Happy Camp, CA: Naturegraph, 1988. $15.95; $8.95pa. ISBN 0911010572; 0911010564pa.

    120 species; color photos; includes place for notes. Revised edition of *Wildflowers of the Redwood Empire*, 3rd ed. Earlier editions published as *Redwood Empire Wildflower Jewels*.

707.   Young, Robert G., and Joann W. Young. **Colorado West: Land of Geology and Wildflowers**. Rev. ed. Grand Junction, CO: R. G. Young, 1984. 239p. $5.50pa. ISBN 0961101008pa.

    200 species; key; color photos; arranged by habitat. About one-half of the book covers geology. Revised edition of *Geology and Wildflowers of Grand Mesa, Colorado*.

Chapter 9

Animals

Like the chapter titled "Plants," this is a catchall section covering field guides with multiple animal types, or animals that do not fit in the other chapters. Land snails, for instance, are included here, while marine and freshwater snails are found in Chapter 10, "Marine Invertebrates."

Terminology can be a problem with some animal groups. "Animal" and "wildlife" are sometimes used as synonyms for "mammal." On the other hand, "wildlife" may mean mammals, all wild animals, or in Britain, wild living things whether plants or animals. In casual usage, connoisseurs often lump amphibians and reptiles together as "herps" (from the Greek *herpetos*, creeping animals).

Many of the animal guides in this chapter are marine guides that cover both vertebrate and invertebrate animals. A number of guides to tracks and signs of mammals and birds are found here, and more are covered in Chapter 15, "Mammals." Also included in this chapter are selected field guide-style books that feature reconstructions of extinct animal species, especially dinosaurs. Field guides for the actual fossils themselves are in Chapter 17, "Geology and Fossils."

See Chapter 2, "Flora and Fauna," for other general animal guides.

NORTH AMERICA

708. Amos, Stephen H. **Familiar Seashore Creatures**. New York: Knopf, 1990. 192p. (Audubon Society Pocket Guides). $6.00pa. ISBN 0679729828pa.

80 species; color photos; arranged by family. Covers both Atlantic and Pacific coasts.

709. Beletsky, Les. **Ecotravellers' Wildlife Guide to Tropical Mexico**. San Diego, CA: Academic, 1999. 300p. (Ecotravellers' Guides). $27.95pa. ISBN 0120848120pa.

500 species; color illustrations and photos. A traveler's guide with natural history and site guides. Not seen; see series entry 33.

710. Boschung, Herbert T., David K. Caldwell, Melba C. Caldwell, Daniel W. Gotshall, and James D. Williams. **The Audubon Society Field Guide to North American Fishes, Whales, and Dolphins**. New York: Knopf, 1983. 848p. (Audubon Society Field Guide Series). $19.00 (flexi). ISBN 0394534050 (flexi).

529 species in detail, another 400 brief descriptions; color photos, some black-and-white illustrations.

711. Brodie, Edmund D., Jr. **Venomous Animals: 300 Animals in Full Color**. New York: Golden, 1989. 160p. (Golden Guide). $5.95pa. ISBN 0307240746 (Scholar's ed.); 0307240746pa.

Color illustrations of all groups from cnidarians to mammals; worldwide. Illustrated by John D. Dawson.

712. Cunningham, Patricia, and Paul Goetz. **Pisces Guide to Venomous and Toxic Marine Life of the World**. Houston, TX: Pisces, 1996. 152p. $18.95pa. ISBN 1559920882pa.

200 species; color photos; includes description of emergency situations and first aid.

713. Elman, Robert. **The Hunter's Field Guide to the Game Birds and Animals of North America**. Rev. ed. New York: Knopf, 1982. 655p. ISBN 0394712609pa.

70 species of game birds, 32 species of mammals; black-and-white photos and line drawings, some color photos; extensive natural history and hunting tactics, as well as tracks and signs.

714. Fichter, George S. **Endangered Animals: 140 Species in Full Color**. New York: Golden, 1995. 160p. (A Golden Guide). $6.00pa. ISBN 0307245012pa; 0307645010.

140 species of animals worldwide; color illustrations. Illustrated by Kristin Kest.

715. Gaffney, Eugene S. **Dinosaurs**. New York: Golden, 1990. 160p. (Golden Guide). $4.50pa. ISBN 0307240762pa.

50 types; color illustrations of reconstructions; natural history. Illustrated by John D. Dawson.

716. Gilliland, Hap. **Mystery Tracks in the Snow: A Guide to Animal Tracks**. Happy Camp, CA: Naturegraph, 1990. 142p. $7.95pa. ISBN 0879611995pa; 0879611987.

   52 mammals, plus a few birds, amphibians, and reptiles; numerous black-and-white illustrations and photos of tracks in various conditions (old, new, snow, sand, etc.). Also includes scenarios showing how to draw conclusions about events/behavior from tracks.

717. Greenberg, Idaz. **Guide to Corals and Fishes of Florida, the Bahamas and the Caribbean**. Miami: Seahawk, 1986. 64p. $7.95pa; $13.95 (waterproof ed.). ISBN 0913008087pa; 0913008079 (waterproof ed.).

   50 species of corals, 210 of fishes, plus a few turtles and dangerous animals; color photos; available in waterproof and standard paperback editions. Both editions are also available in Spanish as *Guía de Corales y Peces* and *Impermeable Guía de Corales y Peces*.

718. Hoffmeister, Donald Frederick. **Zoo Animals**. New York: Golden, 1967. 160p. (A Golden Nature Guide). ISBN 0307635384pa.

   400 species of amphibians and reptiles, birds, and mammals commonly found in zoos; color illustrations. Illustrated by Arthur Singer.

719. Humann, Paul, and Ned DeLoach. **Snorkeling Guide to Marine Life: Florida, Caribbean, Bahamas**. Jacksonville, FL: New World, 1995. 80p. $12.95pa. ISBN 1878348108pa.

   260 species of plants, invertebrates, and fishes; color photos; includes animals found to 15 feet deep; spiralbound.

720. Jahn, Theodore Louis, and Eugene C. Bovee. **How to Know the Protozoa**. 2nd ed. Dubuque, IA: W. C. Brown, 1979. 279p. (Pictured Key Nature Series). $25.80pa. ISBN 069704758X; 0697047598pa.

   375 species; key throughout; black-and-white illustrations; microscope required.

721. Kaplan, Eugene H. **A Field Guide to Coral Reefs of the Caribbean and Florida: A Guide to the Common Invertebrates and Fishes of Bermuda, the Bahamas, Southern Florida, the West Indies, and the Caribbean Coast of Central and South America**. Boston: Houghton Mifflin, 1982. 289p. (Peterson Field Guide Series, No. 27). $19.45; $18.00pa. ISBN 0395316618; 0395469392pa.

   500 species; black-and-white and color photos and line drawings. Illustrated by Susan L. Kaplan.

722. Knutson, Roger M. **Furtive Fauna: A Field Guide to the Creatures Who Live on You**. Berkeley, CA: Ten Speed, 1996. 96p. $7.95pa. ISBN 0898158273pa.

    17 types of animals (mostly arthropods) that live on humans; black-and-white illustrations. Not a true field guide since the emphasis is on (un)natural history rather than identification, but the illustrations are good enough to allow the worried user to distinguish between a flea and a head louse. Originally published in 1992 by Penguin.

723. Kricher, John C. **Peterson First Guide to Dinosaurs**. Boston: Houghton Mifflin, 1990. 128p. $4.95pa. ISBN 0395524407pa.

    100 species; color illustrations. Illustrated by Gordon Morrison.

724. Levy, Charles Kingsley. **A Field Guide to Dangerous Animals of North America, Including Central America**. Brattleboro, VT: Stephen Greene, 1983. 164p. ISBN 0828905037pa.

    100 species of mammals, insects, fishes, arachnids, and reptiles; black-and-white illustrations and color photos; arranged by taxonomic group and type of injury. Also published in Canada as *Guide des Animaux Dangereux de l'Amerique du Nord*. Illustrated by Laszlo Meszoly.

725. Reader's Digest Editors. **North American Wildlife: Mammals, Reptiles, and Amphibians**. New York: Reader's Digest, 1998. 191p. $16.95pa. ISBN 0762100354pa.

    230 species of mammals, 300 reptiles, and 200 amphibians; color illustrations; maps.

726. Robinson, Michael H., David Challinor, with Holly Webber. **Zoo Animals**. New York: Macmillan USA, 1995. 256p. (Smithsonian Guides). $18.00pa; $24.95. ISBN 0028604075pa; 0028604067.

    250 species commonly found in zoos; color photos; natural history; includes maps of 18 popular zoos in the U.S.

727. Rue, Leonard Lee, III. **Complete Guide to Game Animals: A Field Book of North American Species**. 2nd rev. ed. New York: Outdoor Life, 1981. 638p. ISBN 0442277962.

    43 species; black-and-white photos; maps; extensive natural history, trophy size. Less information on hunting strategy than Elman (see entry 713). Previous edition published as *Sportsman's Guide to Game Animals*.

728. Schell, Stewart C. **How to Know the Trematodes**. Dubuque, IA: W. C. Brown, 1970. 355p. (Pictured Key Nature Series).
Comprehensive; keys; line drawings.

729. Schmidt, Gerald D. **How to Know the Tapeworms**. Dubuque, IA: W. C. Brown, 1970. 266p. (Pictured Key Nature Series). ISBN 0697048608.
Covers all genera worldwide; keys; black-and-white illustrations. Requires microscope for identification.

730. Smith, Richard P. **Animal Tracks and Signs of North America**. Harrisburg, PA: Stackpole, 1982. 271p. $16.95pa. ISBN 0811721248pa.
150 species; black-and-white photos of tracks and animals; extensive natural history.

731. Wallace, Joseph E. **Familiar Dinosaurs**. New York: Knopf, 1993. 191p. (National Audubon Society Pocket Guides). $8.00pa. ISBN 067974150Xpa.
78 species; color illustrations plus silhouettes of skeletons.

732. **Wildlife Identification Pocket Guide: With Field Dressing Commentary**. Seattle, WA: Outdoor Empire, 1992. 118p. $2.50pa. ISBN 0916682498pa.
114 species; color photos; includes tracks and scats; divided into large mammals, small mammals, upland birds, waterfowl, non-game birds, and raptors. Also includes the American alligator. Has field dressing information for a generic large and small mammals. For hunters.

Eastern North America

733. Burch, John Bayard. **How to Know the Eastern Land Snails: Pictured-Key for Determining the Land Snails of the United States Occurring East of the Rocky Mountain Divide**. Dubuque, IA: W. C. Brown, 1962. 214p. (Pictured Key Nature Series).
390 species; key; black-and-white illustrations.

734. Carlton, Mike, and John Netherton. **Great Smoky Mountains National Park: Wildlife Watcher's Guide**. Minocqua, WI: NorthWord, 1996. 96p. $11.95pa. ISBN 1559715448pa.
17 mammals, 8 amphibians, 3 reptiles, and 18 birds; color photos; descriptions include tracks, natural history, and wildlife watching tips.

735. Claridge, Edward, and Betty Ann Milligan. **Animal Signatures**. 4th rev. and redesigned ed. Halifax, N.S.: Nimbus, 1992. 72p. (Field Guide Series). $6.95pa. ISBN 1551090481pa.

Covers tracks and signs of Nova Scotia animals, including 20 mammals and 5 birds.

736. Collins, Henry Hill, Jr. **The Complete Field Guide to North American Wildlife, Eastern Edition**. New York: Harper and Row, 1981. 714p. ISBN 0061811637.

1,500 species of marine invertebrates, fresh and saltwater fishes, reptiles, amphibians, birds, and mammals; color and black-and-white illustrations; life lists.

737. DeLorme Publishing. **Wildlife Signatures: A Guide to the Identification of Tracks and Scat**. Freeport, ME: DeLorme, 1983. 47p. (A Maine Geographic book). ISBN 0899330649pa.

30 species of mammals, 4 birds; black-and-white illustrations of animals, tracks, and scats; natural history; pamphlet. Covers Maine mammals. The text is taken largely from *Field Guide Edition of Wild Mammals of New England* by Alfred J. Godin.

738. Elliott, Lang. **Guide to Night Sounds**. Ithaca, NY: NatureSound Studio, 1992. Sound cassette or compact disk and booklet. $13.95 (cassette); $16.95 (CD). ISBN 1878194232 (cassette); 1878194224 (CD).

60 species; includes birds, frogs, insects, and mammals.

739. Elliott, Lang, and Ted Mack. **Wild Sounds of the Northwoods**. Ithaca, NY: NatureSound Studio, 1990. Sound cassette or compact disk and booklet. $12.95 (cassette); $16.95 (CD). ISBN 1878194267 (cassette); 1878194275 (CD).

100 species; includes birds, frogs, insects, and mammals.

740. Headstrom, Richard. **Identifying Animal Tracks: Mammals, Birds, and Other Animals of the Eastern United States**. New York: Dover, 1983. 141p. $4.95pa. ISBN 0486244423pa.

100 species of animals; keys; line drawings. Reprint, originally titled *Whose Track Is It?*

741. Martinez, Andrew J. **Marine Life of the North Atlantic: Canada to New England**. Wenham, MA: Marine Life, 1994. 272p. $32.95pa. ISBN 096401310Xpa.

165 species of marine invertebrates and 35 fishes; color photos. Descriptions include habitat, range, natural history, and space to record date and location seen. Spiralbound.

742. Milstein, Michael. **Badlands, Theodore Roosevelt and Wind Cave National Parks**. Minocqua, WI: NorthWord, 1996. 96p. (Wildlife Watcher's Guide). $9.95pa. ISBN 1559715758pa.

    50 species of birds and mammals; color photos; natural history.

743. Pollock, Leland W. **A Practical Guide to the Marine Animals of Northeastern North America**. New Brunswick, NJ: Rutgers University Press, 1997. 367p. ISBN 0813523982; 0813523990pa.

    1,500 species; tabular keys; black-and-white illustrations. An oversized student's guide to marine invertebrates and vertebrates from zooplankton to marine mammals (but no birds).

Western North America

744. Balazs, George H. **Hawaii's Seabirds, Turtles and Seals**. Honolulu: Worldwide Distributors, 1976. 31p.

    15 birds, 2 turtles, and the Hawaiian monk seal; color photos; natural history; pamphlet. A tourist's guide.

745. Beletsky, Les. **Ecotravellers' Wildlife Guide to Hawaii**. San Diego, CA: Academic, 1999. 350p. (Ecotravellers' Wildlife Guides). $27.95pa. ISBN 0120848139pa.

    Not seen. 300 species; color illustrations and photos. A traveler's guide with natural history and site guides. See series entry 33.

746. Gotshall, Daniel. **Marine Animals of Baja California: A Guide to the Common Fish and Invertebrates**. 2nd ed. Monterey, CA: Sea Challengers, 1987. 113p. $20.95pa. ISBN 0930118154pa.

    140 species of fish, 62 species of invertebrates; color photos. Descriptions include size, habitat, and range.

747. Halfpenny, James C. **Scats and Tracks of the Rocky Mountains**. Helena, MT: Falcon, 1998. 184p. $9.95pa. ISBN 1560446285pa.

    70 species, including 20 birds and reptiles; key; black-and-white illustration of tracks and animals; maps. Has detailed descriptions of animals, tracks, and scats. Illustrated by Todd Telander.

748. Hancock, David, and Susan Hancock. **Guide to the Wildlife of the Rockies**. Saanichton, B.C.: Hancock House, 1975. 33p. $3.50pa. ISBN 0919654339pa.

    4 amphibians, 31 birds, 45 mammals; black-and-white and color photos, black-and-white illustrations; natural history with very little description. Pamphlet.

749. Hinton, Sam. **Seashore Life of Southern California: An Intro-duction to the Animal Life of California Beaches South of Santa Barbara**. Rev. ed. Berkeley, CA: University of California Press, 1987. 217p. (California Natural History Guides, No. 26). $14.95pa. ISBN 0520059247pa; 0520059239.

 300 species of marine animals; black-and-white illustrations and color photos; natural history.

750. Kessler, Doyne W. **Alaska's Saltwater Fishes and Other Sea Life: A Field Guide**. Anchorage, AK: Alaska Northwest, 1985. 358p. ISBN 0882403028pa.

 125 species of fish, 200 of invertebrates; line drawings and color photos.

751. McLachlan, Dan H., and Jak Ayres. **Fieldbook of Pacific Northwest Sea Creatures**. Happy Camp, CA: Naturegraph, 1979. 208p. $12.95pa. ISBN 0879610697; 0879610689pa.

 700 species of animals from sponges to fish; brief field key; black-and-white illustrations.

752. Parsons, Christina. **Dangerous Marine Animals of the Pacific Coast**. San Luis Obispo, CA: Helm; Monterey, CA: Sea Challengers, 1986. 96p. $4.95pa (Helm). ISBN 0936940034pa (Helm); 0930118111pa (Sea Challengers).

 35 species or groups of animals; black-and-white illustra-tions; includes description, type of hazard, prevention, symptoms, and first aid. Illustrated by Ron Stark and Shira Stark.

753. Ransom, Jay Ellis. **Harper and Row's Complete Field Guide to North American Wildlife, Western Edition**. New York: Harper and Row, 1981. 809p. ISBN 0061817155.

 1,800 species of marine invertebrates, fishes, amphibians, reptiles, birds, and mammals; color and black-and-white illustra-tions separate from text.

754. Wilkinson, Todd. **Glacier Park Wildlife: A Watcher's Guide: Includes Listings for Waterton Lakes National Park**. Minocqua, WI: NorthWord, 1993. 96p. $11.95pa. ISBN 1559711442pa.

 30 mammals, 13 birds; color photos; natural history.

755. Wilkinson, Todd. **Rocky Mountain National Park: A Wildlife Watcher's Guide**. Minocqua, WI: NorthWord, 1994. 96p. $11.95pa. ISBN 1559712279pa.

 26 species of mammals and 18 birds often seen from road-sides; color photos, tracks; natural history; sites.

756. Wilkinson, Todd. **Yellowstone Wildlife: A Watcher's Guide**. Minocqua, WI: NorthWord, 1992. $11.95pa. ISBN 155971140Xpa.

   14 birds and 31 mammals; color photos, tracks; natural history; sites for viewing wildlife.

757. Wilkinson, Todd. **Grand Canyon, Zion, Bryce National Parks: A Wildlife Watcher's Guide**. Minocqua, WI: NorthWord, 1995. 144p. (A Wildlife Watcher's Guide). $14.95pa. ISBN 1559714611pa.

   22 mammals, 3 amphibians, 6 reptiles; 18 birds; color photos; natural history.

Marine
Invertebrates

Many of the field guides found in this chapter deal with seashells. There are fewer guides to living molluscs, but they do exist, along with guides to other marine invertebrates such as nudibranchs, corals, and crabs. Field guides to freshwater invertebrates, primarily molluscs such as mussels and snails, are also included in this chapter. For more marine and freshwater invertebrate guides, see Chapter 2, "Flora and Fauna," and Chapter 9, "Animals."

NORTH AMERICA

757a. Abbott, R. Tucker. **Seashells**. New York: Gallery, 1990. 176p. (American Nature Guides). ISBN 083176967X.

400 species; color photos; northern hemisphere. Published in England as *Seashells of the Northern Hemisphere*.

758. Abbott, R. Tucker. **Seashells of North America: A Guide to Field Identification**. New York: Golden, 1968. 280p. (A Golden Field Guide). $11.95pa. ISBN 0307636577; 0307470032; 0307136574pa.

850 species; color illustrations. Illustrated by George F. Sandstrom.

759. Abbott, R. Tucker. **Seashells of the World: A Guide to the Better-known Species**. Rev. ed. New York: Golden, 1985. 160p. (Golden Guide). $5.95pa. ISBN 0307635147; 0307244105pa.

500 species; color illustrations. Also published in Mexico as *Conchas Marinas del Mundo*. Illustrated by George F. and Marita Sandstrom.

760. Dance, S. Peter. **Shells**. 1st American ed. New York: Dorling Kindersley, 1992. 256p. (Eyewitness Handbooks). $29.95; $17.95 (flexi). ISBN 1564580326; 1564580601 (flexi).

    500 species worldwide; color photos.

761. Douglass, Jackie Leatherbury. **Peterson First Guide to Shells of North America**. Boston: Houghton Mifflin, 1989. 128p. $4.95pa. ISBN 0395482976pa.

    200 species; color illustrations. Illustrated by John Douglass.

762. Fitzpatrick, Joseph F. **How to Know the Freshwater Crustacea**. Dubuque, IA: W. C. Brown, 1983. 227p. (Pictured Key Nature Series). $25.80pa. ISBN 0697047830pa.

    Key throughout; black-and-white illustrations. Technical language; intended for use of students or researchers. Covers most genera of the Americas; most genera are keyed to species level.

763. Forey, Pamela, and Cecilia Fitzsimons. **An Instant Guide to Seashells: The Most Familiar Species of North American Seashells Described and Illustrated in Color**. New York: Bonanza, 1987. 124p. ISBN 0517635488.

    143 species; color illustrations; maps; checklist. Arranged by form of shell.

764. Meinkoth, Norman August. **The Audubon Society Field Guide to North American Seashore Creatures**. New York: Knopf, 1981. 799p. (Audubon Society Field Guide Series). $19.00 (flexi). ISBN 0394519930 (flexi).

    666 species covered in detail, another 220 in brief; color photos.

765. Rehder, Harald A. **The Audubon Society Field Guide to North American Seashells**. New York: Knopf, 1981. 894p. (Audubon Society Field Guide Series). $19.00 (flexi). ISBN 0394519132 (flexi).

    671 species covered in detail, another 200 mentioned; color photos.

766. Rehder, Harald A. **Familiar Seashells**. New York: Knopf, 1987. 192p. (Audubon Society Pocket Guides). $9.00pa. ISBN 0394757955pa.

    80 species of gastropods, chitons, and bivalves; color photos; covers Pacific, Atlantic, and Gulf coasts of North America.

767. Romashko, Sandra. **The Complete Collector's Guide to Shells and Shelling: Sea Shells from the Waters of the North American Atlantic and Pacific Oceans, Gulf of Mexico, Gulf of California,**

the Caribbean, the Bahamas, and Hawaii. 2nd ed. Miami, FL: Windward, 1994. 112p. $9.95pa. ISBN 0893170321pa.

585 species; color photos.

768. Romashko, Sandra. **The Shell Book: The Complete Guide to Collecting and Identifying with a Special Section on Starfish and Other Sea Creatures**. 2nd ed. Miami, FL: Windward, 1974. 64p. $5.95pa. ISBN 0684138387pa.

300 species; color photos. On cover: *The Shell Book: Atlantic, Gulf, and Caribbean.*

769. Sabelli, Bruno. **Simon and Schuster's Guide to Shells**. New York: Simon and Schuster, 1980. 512p. $12.95pa. ISBN 0671253190; 0671253204pa.

357 species; color and black-and-white photos; arranged by habitat; includes descriptions of shells and bodies; worldwide. Translation of *Conchiglie*. Also published as *The Macdonald Encyclopedia of Shells.*

Eastern North America

770. Abbott, R. Tucker. **Collectible Florida Shells**. Melbourne, FL: American Malacologists, 1984. 64p. (Collectible Shells of Southeastern U.S., Bahamas and Caribbean). $5.95pa. ISBN 0915826119pa; 0915826127 (waterproof ed.).

200 species; color photos; pamphlet. Waterproof edition printed on plastic sheets.

771. Abbott, R. Tucker. **Collectible Shells of Southeastern U.S., Bahamas and Caribbean**. Melbourne, FL: American Malacologists, 1984. 64p. ISBN 0915826143pa; 0915826135 (waterproof).

200 species; color photos; pamphlet. Waterproof edition printed on plastic sheets.

772. Abbott, R. Tucker, and Percy A. Morris. **A Field Guide to Shells of the Atlantic and Gulf Coasts and the West Indies**. 4th ed. Boston: Houghton Mifflin, 1995. 350p. (Peterson Field Guide Series, No. 3). $16.95pa. ISBN 0395697808; 0395697794pa.

800 species; color photos, black-and-white illustrations; descriptions include range, habitat, and remarks; list of shell clubs. Includes cephalopods. First published in 1947 as *A Field Guide to the Shells of our Atlantic Coast* by Morris.

773. Andrews, Jean. **A Field Guide to Shells of the Florida Coast**. Houston, TX: Gulf, 1994. 182p. $16.95pa. ISBN 0877192499; 0877192340pa.

    270 species; black-and-white and color photos; arranged taxonomically.

774. Andrews, Jean. **A Field Guide to Shells of the Texas Coast**. Houston, TX: Gulf, 1992. 176p. (Texas Monthly Field Guide Series). $23.95; $16.95pa. ISBN 0877192111; 0877192103pa.

    279 species; black-and-white and some color photos. Originally published as *Texas Shells: A Field Guide* in 1981.

775. Berrick, Stephan. **Crabs of Cape Cod**. Brewster, MA: Cape Cod Museum of Natural History, 1986. 77p. (Natural History Series, No. 3). $6.95pa. ISBN 0916275019; 0916275001pa.

    20 species; key; line drawings; natural history. Illustrated by Terry Ellis.

776. Cummings, Kevin S., and Christine A. Mayer. **Field Guide to Freshwater Mussels of the Midwest**. Champaign, IL: Illinois Natural History Survey, 1992. 194p. $15.00. (Illinois Natural History Survey Manual, No. 5).

    75 species; brief key; color photos; maps.

777. Gordon, Julius, and Townsend Weeks. **Seashells of the Northeast Coast, from Cape Hatteras to Newfoundland**. Blaine, WA: Hancock House, 1982. 64p. ISBN 0888390807.

    125 species; color photos.

778. Greenberg, Margaret H., and Nancy J. Olds. **The Sanibel Shell Guide**. Winter Park, FL: Anna, 1982. 117p. $5.95pa. ISBN 0893050415pa.

    100 species; black-and-white and color photos; includes information on where and when to find shells at Sanibel Island, Florida.

779. Howells, Robert G., Raymond W. Neck, and Harold D. Murray. **Freshwater Mussels of Texas**. Austin, TX: Texas Parks and Wildlife Dept., Inland Fisheries Division: Distributed by University of Texas Press, 1996. 218p. (Learn about Texas). $29.95pa. ISBN 1885696108pa.

    52 species; color and black-and-white photos; maps; natural history and background information, including use of mussels. This oversized volume uses technical language.

780. Huggins, Donald, Paul M. Liechti, Leonard C. Ferrington, Jr., eds. **Guide to the Freshwater Invertebrates of the Midwest**. 2nd ed. Lawrence: Kansas Biological Survey, 1985. 221p. (Technical Publications, Kansas Biological Survey, The University of Kansas, No. 11).

    Common species; keys throughout; black-and-white illustrations and photos; includes lists of endangered and threatened invertebrates in the area and information on studying aquatic environments. Covers Iowa, Kansas, Missouri, and Nebraska.

781. Humann, Paul. **Reef Coral Identification: Florida, Caribbean, Bahamas, Including Marine Plants**. Rev. 2nd printing. Jacksonville, FL: New World, 1993. 239p. $31.50pa. ISBN 1878348035pa.

    250 species; color photos. Spiralbound.

782. Humann, Paul, and Ned Deloach. **Reef Creature Identification: Florida, Caribbean, Bahamas**. Jacksonville, FL: New World 1992. 320p. $37.95pa. ISBN 1878348019pa.

    450 species; color photos; checklist; spiralbound. Second of a three-volume set on reef identification. (Volume 1, *Reef Fish Identification*, volume 3, *Reef Coral Identification*.)

783. Lipe, Robert E., and R. Tucker Abbott. **Living Shells of the Caribbean and Florida Keys**. Melbourne, FL: American Malacologists, 1991. 80p. ISBN 0915826259pa.

    390 species; color photos; attractive pamphlet.

784. Ruppert, Edward E., and Richard Fox. **Seashore Animals of the Southeast: A Guide to Common Shallow-Water Invertebrates of the Southeastern Atlantic Coast**. Columbia: University of South Carolina Press, 1988. 429p. $29.95pa. ISBN 0872495345; 0872495353pa.

    740 common species discussed, 360 illustrated; visual key; line drawings and color and black-and-white photos. Covers invertebrates found at wading depths.

785. Smith, Frederick George Walton. **Atlantic Reef Corals: A Handbook of the Common Reef and Shallow-Water Corals of Bermuda, the Bahamas, Florida, the West Indies, and Brazil**. Rev. ed. Coral Gables, FL: University of Miami Press, 1976. 164p. $15.95. ISBN 0870241796.

    60 species; keys; black-and-white photos; checklist; background information.

786. Thompson, Fred G. **The Freshwater Snails of Florida: A Manual for Identification**. Gainesville: University Presses of Florida, 1984. 94p. ISBN 081300781Xpa.

    Comprehensive; keys; black-and-white illustrations. Illustrated by Erland Larson.

787. Zinn, Donald J. **Marine Mollusks of Cape Cod**. Brewster, MA: Cape Cod Museum of Natural History, 1984. 78p. (Natural History Series, No. 2). $6.95pa. ISBN 0916275000pa.

    70 species; line drawings; natural history; includes 28 recipes. Illustrated by Terry Ellis.

Western North America

788. Barr, Lou, and Nancy Barr. **Under Alaskan Seas: The Shallow Water Marine Invertebrates**. Anchorage, Alaska: Alaska Northwest, 1983. 208p. ISBN 0882402358pa.

    241 species; color photos and black-and-white illustrations; covers the most common intertidal and shallow-water species. Fairly technical language.

789. Behrens, David W. **Pacific Coast Nudibranchs: A Guide to the Opisthobranchs, Alaska to Baja California**. 2nd ed. Monterey, CA: Sea Challengers, 1991. 107p. $21.95pa. ISBN 0930118170pa.

    217 species; keys; color photos; oversized.

790. Bertsch, Hans, and Scott Johnson. **Hawaiian Nudibranchs: A Guide for Scuba Divers, Snorkelers, Tidepoolers, and Aquarists**. Honolulu: Oriental, 1981. 112p. $15.95pa. ISBN 0932596150pa.

    90 species; color photos. Includes information on range and food size of each species.

791. Brusca, Richard C. **Common Intertidal Invertebrates of the Gulf of California**. 2nd ed., revised and expanded. Tucson: University of Arizona Press, 1980. 513p. ISBN 0816517541; 0816506825pa.

    1,300 species; field manual; keys; black-and-white photos and drawings. Revised edition of *A Handbook to the Common Intertidal Invertebrates of the Gulf of California*.

792. Fielding, Ann. **Hawaiian Reefs and Tidepools: A Guide to Hawaii's Shallow-Water Invertebrates**. 2nd ed. Honolulu: Oriental, 1982. 103p. ISBN 0932596118.

    150 species; color photos; natural history; a tourist guide.

793. Foster, Nora Rakestraw. **Intertidal Bivalves: A Guide to the Common Marine Bivalves of Alaska**. Fairbanks: University of Alaska Press, 1991. 152p. $30.00; $20.00pa. ISBN 0912006498; 0912006544pa.

     106 species; keys; black-and-white illustrations; includes distribution and habitat. Oversized.

794. Gosliner, Terrence, David W. Behrens, and Gary C. Williams. **Coral Reef Animals of the Indo-Pacific: Animal Life from Africa to Hawaii Exclusive of the Vertebrates**. Monterey, CA: Sea Challengers, 1996. 314p. $45.00pa. ISBN 0930118219pa.

     1,150 species; visual key to major groups; color photos; descriptions include identification, natural history, and distribution. Described as a "scientific field guide" for divers, aquarists, and biologists.

795. Gotshall, Daniel. **Guide to Marine Invertebrates: Alaska to Baja California**. Monterey, CA: Sea Challengers, 1994. 105p. $22.95pa. ISBN 0930118197pa.

     254 species of common subtidal invertebrates; visual key; color photos; natural history.

796. Gotshall, Daniel, and Laurence L. Laurent. **Pacific Coast Subtidal Marine Invertebrates: A Fishwatcher's Guide**. Los Osos, CA: Sea Challengers, 1979. 107p. $16.95pa. ISBN 0930118022pa; 0930118030.

     161 species; pictorial key; color photos.

797. Harbo, Rick M. **Shells and Shellfish of the Pacific Northwest: A Field Guide**. Madiera Park, B.C.: Harbour, 1997. 270p. $26.95pa. ISBN 1550171461pa.

     225 common molluscs; color photos; guide to bivalves by siphon or "show" (portion of mollusc showing above the ground); checklist. Photos and brief descriptions in front, followed by species accounts.

798. Jensen, Gregory C. **Pacific Coast Crabs and Shrimps**. Monterey, CA: Sea Challengers, 1995. 87p. $21.95pa. ISBN 093011820pa.

     163 species; color photos; checklist. Descriptions include range, habitat, and natural history.

799. Kay, E. Alison, and Olive Schoenberg-Dole. **Shells of Hawai'i**. Honolulu: University of Hawaii Press, 1991. 89p. $12.95pa. ISBN 0824813162pa.

     300 species; color photos; background information; no description of pictured shells.

800.   Kerstitch, Alex N. **Sea of Cortez Marine Invertebrates: A Guide for the Pacific Coast, Mexico to Ecuador**. Monterey, CA: Sea Challengers, 1989. 120p. $20.95pa. ISBN 0930118146pa.

   283 species; color photos. Descriptions include size, habitat, and distribution.

801.   Morris, Percy A. **A Field Guide to Pacific Coast Shells, Including Shells of Hawaii and the Gulf of California**. 2nd ed., rev. and enl. Boston: Houghton Mifflin, 1974. 297p. (Peterson Field Guide Series, No. 6). $21.95; $15.95pa. ISBN 0395080290; 0395183227pa.

   945 common species; black-and-white and color photos. Previously published as *A Field Guide to Shells of the Pacific Coast and Hawaii*.

802.   Russo, Ron, and Pam Olhausen. **Pacific Intertidal Life: A Guide to Organisms of Rocky Reefs and Tide Pools of the Pacific Coast**. Berkeley, CA: Nature Study Guild, 1981. $3.00pa. ISBN 0912550104pa.

   79 species of animals, 15 of algae; key throughout; black-and-white illustrations; pocket sized.

803.   Wrobel, David, and Claudia Mills. **Pacific Coast Pelagic Invertebrates: A Guide to the Common Gelatinous Animals**. Monterey, CA: Sea Challengers, Monterey Bay Aquarium, 1998. 108p. $16.95pa. ISBN 0930118235pa.

   180 species; color photos; descriptions include natural history and distribution and habitat. Arranged in taxonomic order.

Chapter 11

Insects and Arachnids

Insects and their cousins, the arachnids (spiders, scorpions, ticks, and mites), are among the most numerous groups of animals on earth. Only a few insect or arachnid orders are identifiable to the species level by amateurs (or are of interest), so most field guides either identify the common orders and discuss a few typical species, or cover only a single order such as butterflies or spiders. Harmful or beneficial insects found around the home or garden are also common field guide subjects. Many butterfly books identify caterpillars and provide information on larval food plants, but other larval insects are rarely illustrated or described.

Other insect field guides are found in Chapter 2, "Flora and Fauna," and Chapter 9, "Animals."

NORTH AMERICA

804. Arnett, Ross H., Jr., and Richard L. Jacques, Jr. **Simon and Schuster's Guide to Insects**. New York: Simon and Schuster, 1981. 512p. (A Fireside Book). $15.00pa. ISBN 0671250140pa; 0671250132.

    350 species worldwide; color photos; no spiders.

805. Arnett, Ross H., Jr., N. M. Downie, and H. E. Jaques. **How to Know the Beetles**. 2nd ed. Dubuque, IA: W. C. Brown, 1980. 416p. (Pictured Key Nature Series). $25.80pa. ISBN 0697047768, 0697047768pa.

    1,500 species from all families found in the U.S. and Canada, 900 species illustrated; key throughout; black-and-white illustrations.

806. Bland, Roger G., and H. E. Jaques. **How to Know the Insects**. 3rd ed. Dubuque, IA: W. C. Brown, 1978. 409p. (Pictured Key Nature Series). $25.80pa. ISBN 0697047539; 0697047520pa.

Key to families, with select common species included; black-and-white illustrations and line drawings. Also includes information on collecting insects. Second edition by H. E. Jaques.

807. Borror, Donald J., and Richard E. White. **A Field Guide to Insects of America North of Mexico**. Boston: Houghton Mifflin, 1998. 404p. (Peterson Field Guide Series, No. 19). $27.00; $18.00pa. ISBN 0395911710; 0395911702pa.

579 families, each with at least one species illustrated; 1,300 drawings plus color illustrations. Illustrated by the authors.

808. Chu, Hung-fu. **How to Know the Immature Insects**. 2nd ed. _que, IA: William C. Brown, 1992. 352p. (Pictured Key Series). _pa. ISBN 0697048063pa.

Keys to about 400 families worldwide; black-and-white illustrations.

809. Daccordi, Mauro, Paolo Triberti, and Adriano Zanetti. **Simon and Schuster's Guide to Butterflies and Moths**. New York: Simon and Schuster, 1988. 383p. ISBN 0671660667pa.

300 species; color photos; common insects worldwide. Translation of _Farfalle_.

810. Ehrlich, Paul R., and Anne H. Ehrlich. **How to Know the Butterflies: Illustrated Keys for Determining to Species All Butterflies Found in North America, North of Mexico, With Notes on Their Distribution, Habits, and Larval Food, and Suggestions for Collecting and Studying Them**. Dubuque, IA: W. C. Brown, 1961. 262p. (Pictured Key Nature Series).

Comprehensive; key throughout; black-and-white illustrations, including many larvae and pupae.

811. Emerton, James Henry. **The Common Spiders of the United States**. York: Dover, 1961. 227p. $6.95pa. ISBN 0486202232pa.

200 species; line drawings and color photos; checklist.

812. Farrand, John, ed. **Familiar Insects and Spiders**. New York: Knopf, 1988. 192p. (Audubon Society Pocket Guides). $9.00pa. ISBN 0394757920pa.

75 species of insects and 5 species of spiders are illustrated, another 13 mentioned; color photos; includes identification, habitat, range, life cycle.

813.	Feltwell, John. **Butterflies of North America**. New York: Smithmark, 1992. 192p. (American Nature Guides). $9.98. ISBN 0831769637.

	350 species; color illustrations. Descriptions include habitat, flight period, and distribution. Very concise. Illustrated by Brian Hargreaves.

814.	Forey, Pamela, and Cecilia Fitzsimons. **An Instant Guide to Butterflies: The Most Familiar Species of North American Butterflies Described and Illustrated in Color**. New York: Bonanza, 1987. 124p. ISBN 051761801X.

	150 species; color illustrations; arranged by color.

815.	Forey, Pamela, and Cecilia Fitzsimons. **An Instant Guide to Insects: The Most Familiar Species of North American Insects Described and Illustrated in Color**. New York: Bonanza, 1987. 125p. (Instant Guide Series). ISBN 051763547X.

	100 groups; color illustrations.

816.	Hafele, Rick, and Scott Roederer. **An Angler's Guide to Aquatic Insects and Their Imitations**. Rev. ed. Boulder, CO: Johnson, 1995. 192p. $15.95. ISBN 1555661610.

	Keys; color illustrations; information on when to use various species of insects or artificial flies when fishing. Illustrated by Richard Bunse.

817.	Helfer, Jacques R. **How to Know the Grasshoppers, Crickets, Cockroaches, and Their Allies**. New York: Dover, 1987. 363p. $10.95pa. ISBN 0486253953pa.

	Comprehensive; key; line drawings. Illustrated by the author.

818.	Kaston, Benjamin Julian. **How to Know the Spiders**. 3rd ed. Dubuque, IA: W. C. Brown, 1978. 272p. (Pictured Key Nature Series). $25.80pa. ISBN 0697048985pa.

	Common species; key throughout; black-and-white illustrations; introductory material.

819.	Leahy, Christopher W. **Peterson First Guide to Insects of North America**. Boston: Houghton Mifflin, 1987. 128p. $4.95pa. ISBN 0395356407pa.

	200 species or families; color illustrations. Illustrated by Richard E. White.

820.   Lehmkuhl, Dennis M. **How to Know the Aquatic Insects**. Dubuque, IA: W. C. Brown, 1979. 168p. (Pictured Key Nature Series). $29.95pa. ISBN 0697047660; 0697047679pa.
Key to families or genus; line drawings.

821.   Levi, Herbert Walter, and Lorna R. Levi. **A Guide to Spiders and Their Kin**. New York: Golden, 1990. 160p. (Golden Guide). $6.00pa. ISBN 0307240215pa.
About 300 common species, mostly North American; color illustrations. Illustrated by Nicholas Strekalovsky.

McDaniel, Burruss. **How to Know the Mites and Ticks**. Dubuque, IA: Wm. C. Brown, 1979. 335p. (Pictured Key Nature Series). ISBN 0697047563; 0697047571pa.
Key to genus, many species included; line drawings. Requires microscope or hand lens.

823.   McGavin, George C. **Insects**. New York: Smithmark, 1992. 208p. (American Nature Guides). $9.98. ISBN 0831769513.
Not seen; see series entry 2.

824.   Miller, Millie, and Cyndi Nelson. **Painted Ladies: Butterflies of North America**. Boulder, CO: Johnson, 1993. 1 v. (unpaged) (Millie and Cyndi's Pocket Nature Guides). $5.95pa. ISBN 1555661033pa.
50 species; color illustrations. Small (4x6 inches); attractively calligraphed.

825.   Milne, Lorus Johnson, and Margery Milne. **The Audubon Society Field Guide to North American Insects and Spiders**. New York: Knopf, 1980. 989p. (Audubon Society Field Guide Series). $19.00 (flexi). ISBN 0394507630 (flexi).
600 species covered in detail, another 250 briefly described; color photos and black-and-white illustrations.

826.   Mitchell, Robert T., and Herbert S. Zim. **Butterflies and Moths: A Guide to the More Common American Species**. New York: Golden, 1964. 160p. (Golden Guide). $14.60; $6.00pa. ISBN 0307640523; 0307240525pa.
423 species; color illustrations; maps. Illustrated by Andre Purenceau.

827.   Opler, Paul A. **Peterson First Guide to Butterflies and Moths**. Boston: Houghton Mifflin, 1994. 128p. $4.95pa. ISBN 0395670721pa.
183 species, color illustrations. Illustrated by Amy Bartlett Wright.

828. Pobst, Dick. **Trout Stream Insects: An Orvis Streamside Guide**. New York: Lyons and Burford, 1990. 81p. $16.95pa. ISBN 1558210679pa.

    46 species illustrated, 15 species of caddisflies described; color photos; arranged by time of hatch and location (eastern and western U.S.). Also includes color photos and descriptions of artificial flies corresponding to most species. A fly fisherman's guide. See also entry 1352 for a guide to artificial flies.

829. Pyle, Robert Michael. **The Audubon Society Field Guide to North American Butterflies**. New York: Knopf, 1981. 916p. (Audubon Society Field Guide Series). $19.00 (flexi). ISBN 0394519140 (flexi).

    600 species covered in detail, another 70 mentioned; color photos; maps; visual key.

830. Schneck, Marcus. **Butterflies: How to Identify and Attract Them to Your Garden**. Emmaus, PA: Rodale, 1990. 160p. $24.95pa. ISBN 0878579176pa.

    250 species; color illustrations; maps; background information; includes host plants and life cycles. A backyard field guide, too large for the hiker.

831. Slater, James Alexander, and Richard M. Baranowski. **How to Know the True Bugs (Hemiptera-Heteroptera)**. Dubuque, IA: W. C. Brown, 1978. 256p. (Pictured Key Nature Series). ISBN 0697048934; 0697048942pa.

    750 common or conspicuous species; key throughout; black-and-white illustrations.

832. Walton, Richard K., and Paul A. Opler. **Familiar Butterflies**. New York: Knopf, 1990. 191p. (Audubon Society Pocket Guides). $9.00pa. ISBN 067972981Xpa.

    80 species illustrated with some similar species mentioned; color photos.

833. White, Richard E. **A Field Guide to the Beetles of North America**. Boston: Houghton Mifflin, 1983. 368p. (Peterson Field Guide Series, No. 29). $21.45; $18.00pa. ISBN 0395318084; 0395339537pa.

    Covers 111 families and over 600 species; visual key to the principle families; black-and-white and color illustrations. Illustrated by the author.

834. Wilsdon, Christina. **National Audubon Society First Field Guide to Insects**. New York: Scholastic, 1998. 160p. $10.95 (flexi); $17.95. ISBN 059005483X (flexi); 0590054473; 0590054848pa.

   50 common species covered in detail, another 125 illustrated and described; color photos; extensive introductory material; children's guide. Includes a separate pocket-sized card illustrating the 50 main species.

835. Wright, Amy Bartlett. **Peterson First Guide to Caterpillars of North America**. Boston: Houghton Mifflin, 1993. 128p. $4.95pa. ISBN 0395564999pa.

   125 species; color illustrations of caterpillars and adults. Illustrated by the author.

836. Zim, Herbert Spencer, and Clarence Cottam. **Insects: A Guide to Familiar American Insects**. Rev. ed. New York: Golden, 1987. 160p. (Golden Guide). $6.00pa. ISBN 030724055Xpa.

   150 families; key; color illustrations; maps. Illustrated by James G. Irving.

Eastern North America

837. Allen, Thomas J. **The Butterflies of West Virginia and Their Caterpillars**. Pittsburgh, PA: University of Pittsburgh Press, 1997. 388p. (Pitt Series in Nature and Natural History). $37.50pa. ISBN 0822956578pa; 0822939738.

   128 species; color photos of adults, pupae, and caterpillars; maps. Descriptions include distribution, habitat, life history, nectar source, and larval host plant.

838. Carpenter, Virginia. **Dragonflies and Damselflies of Cape Cod**. Brewster, MA: Cape Cod Museum of Natural History, 1991. 77p. $10.95pa. ISBN 091627506Xpa.

   62 species described, 41 species illustrated; key; color illustrations; checklist. Illustrated by Terry Ellis.

839. Covell, Charles V. **A Field Guide to the Moths of Eastern North America**. Boston: Houghton Mifflin, 1984. 496p. (Peterson Field Guide Series, No. 30). $21.45; $17.95pa. ISBN 0395260566; 0395361001pa.

   1,300 species; black-and-white and color photos, line drawings. Illustrated by Elaine R. Snyder Hodges and the author.

840. Drees, Bastiaan M., and J. A. Jackman. **Texas Monthly Field Guide to the Common Insects of Texas**. Houston, TX: Gulf, 1998. 300p. (Texas Monthly Field Guide Series). $18.95pa. ISBN 0877192634pa.

   375 common species; color photos and black-and-white illustrations; descriptions include life cycle, pest status, habitat, and food source. Also includes appendixes on collecting insects, endangered insects and arachnids, and a poster showing some of the most common Texas insects.

841. Dunkle, Sidney W. **Damselflies of Florida, Bermuda, and the Bahamas**. Gainesville, FL: Scientific, 1990. 148p. (Scientific Publishers Nature Guide, No. 3). $19.95; $14.95pa. ISBN 0945417861; 0945417853pa.

   46 species; quick key; color photos; natural history.

842. Dunkle, Sidney W. **Dragonflies of the Florida Peninsula, Bermuda, and the Bahamas**. Gainesville, FL: Scientific, 1989. 155p. (Scientific Publishers Nature Guide, No. 1). $14.95pa. ISBN 0945417233pa.

   94 species; color photos; checklist; useful for all of the southeast U.S.

843. Dunn, Gary A. **Insects of the Great Lakes Region**. Ann Arbor: University of Michigan Press, 1996. 324p. (Great Lakes Environment). $42.50; $17.95pa. ISBN 0472095153; 0472065157pa.

   Covers all orders of insects of the region; key to order plus quick keys for each major group; black-and-white illus about 200 species; natural history. Includes appendixes with entomological organizations, insect collections, regulations on collecting insects, and endangered species. More of a manual, but written for the interested amateur.

844. Fichter, George S. **Insect Pests**. New York: Golden, 1987. 160p. (Golden Guide). ISBN 0307240169pa.

   350 species; color illustrations; arranged by type (household, biting, etc.); covers the Midwest region of North America. Illustrated by Nicholas Strekalovsky.

845. Gerberg, Eugene J., and Ross H. Arnett, Jr. **Florida Butterflies**. Baltimore, MD: Natural Science, 1989. 90p. $9.95pa. ISBN 0891400311pa.

   164 species; color photos; checklist; background information.

846.   Glassberg, Jeffrey. **Butterflies Through Binoculars: The East.**
New York: Oxford University Press, 1999. 242p. $18.95pa. ISBN
0195106687pa.

   700 species; color photos; maps; checklist. Expansion of the
author's *Butterflies Through Binoculars: A Field Guide to Butterflies
in the Boston-New York-Washington Region.*

847.   Heitzman, J. Richard, and Joan E. Heitzman. **Butterflies and
Moths of Missouri.** Jefferson City, MO: Missouri Dept. of Conservation, 1987. 385p. $13.50pa.

   300 species; color photos; includes status, food plants, and
larvae.

848.   Jackman, J. A. **A Field Guide to Spiders and Scorpions of
Texas.** Houston, TX: Gulf, 1997. 201p. (Texas Monthly Field
Guide Series). $18.95pa. ISBN 0877192642pa.

   125 species; color photos and black-and-white illustrations;
checklist. Cover title: *Field Guide to Spiders and Scorpions.*

849.   Moulder, Bennett. **A Guide to the Common Spiders of Illinois.**
Springfield, IL: Illinois State Museum, 1992. 125p. (Illinois State
Museum Popular Science Series, No. 10). $10.00pa. ISBN
0897921356pa.

   100 species; key to families; line drawings; checklist; natural
history.

850.   Neck, Raymond. **A Field Guide to the Butterflies of Texas.**
Houston, TX: Gulf, 1996. 298p. (Gulf's Field Guide Series). $21.95pa.
ISBN 087719243Xpa.

   446 species; black-and-white illustrations and color photos;
maps. Descriptions include food plants, life history, range, and
comments.

851.   Opler, Paul A. **A Field Guide to Eastern Butterflies.** 2nd ed. Boston:
Houghton Mifflin, 1998. 486p. (Peterson Field Guide Series, No. 4).
$20.00pa. ISBN 0395904536pa.

   524 species; color illustrations and photos; maps. First edition by Alexander Klots, *A Field Guide to the Butterflies of North
America, East of the Great Plains.* Illustrated by Vichai Malikul.

852.   Royer, Ronald A. **Butterflies of North Dakota: An Atlas and
Guide.** Minot, ND: Minot State University, 1988. 192p. $14.95pa.
ISBN 0961963506pa.

   550 species; keys; color photos; maps. Spiralbound.

853. Tveten, John L., and Gloria A. Tveten. **Butterflies of Houston and Southeast Texas**. Austin: University of Texas Press, 1996. 304p. (Corrie Herring Hooks Series, No. 32). $45.00; $19.95pa. ISBN 0292781423; 0292781431pa.

    100 species; color photos of butterflies and caterpillars; checklist; background information; index of larval and nectar plants. Includes distribution, habitat, host plant. Oversized.

854. Woodbury, Elton N. **Butterflies of Delmarva**. Centreville, MD: Published in association with the Delaware Nature Society by Tidewater, 1994. 138p. $12.95pa. ISBN 0870334530pa.

    61 species; color photos separate from text; includes food sources for both adults and larvae.

855. Yanega, Douglas. **Field Guide to Northeastern Longhorned Beetles (Coleoptera: Cerambycidae)**. Champaign, IL: Illinois Natural History Survey, 1996. (Illinois Natural History Survey Manual, No. 6). $15.00. ISBN 1882932013.

    342 species; color photos. Includes background information on natural history and on collecting specimens. Covers North America west to Minnesota, Iowa, and Missouri, and south to Kentucky and Virginia. Arranged with species accounts (name, flight period, larval feeding habits, size, and notes) followed by plates with brief descriptions. A technical subject, but intended for non-professionals to use in biomonitoring projects.

Western North America

856. Acorn, John Harrison. **Butterflies of Alberta**. Edmonton, Alta.: Lone Pine, 1993. 143p. $13.95pa. ISBN 1551050285pa.

    156 species; color photos; maps; checklist; background information; natural history.

857. Bailowitz, Richard A., and James P. Brock. **Butterflies of Southeastern Arizona**. Tuscon, AZ: Sonoran Arthropod Studies, 1991. 342p. $29.95pa. ISBN 0962662909pa.

    Black-and-white and color photos; species accounts include larval foodplants, flight period, distribution, and comments. Designed for use with another field guide, because the emphasis is not on identification.

858. Bailowitz, Richard A., and Douglas Danforth. **70 Common Butterflies of the Southwest**. Tucson, AZ: Southwest Parks and Monuments Association, 1997. (unpaged). ISBN 1877856843pa.

70 species; color photos; descriptions include size, range, host plants, and flight season.

859. Christensen, James R. **A Field Guide to the Butterflies of the Pacific Northwest**. Moscow, ID: University Press of Idaho, 1981. 116p. (A Northwest Naturalist Book). ISBN 089301074Xpa.
185 species; color photos separate from text; checklist.

860. Garth, John S., and J. W. Tilden. **California Butterflies**. Berkeley, CA: University of California Press, 1985. 246p. (California Natural History Guides, No. 51). $35.00; $14.95pa. ISBN 0520053893pa; 0520052498.
235 species; line drawings and black-and-white and color illustrations separate from text; background information; includes larval plant preferences; checklist. Illustrated by David Mooney and Gene M. Christman.

861. Hogue, C. **Insects of the Los Angeles Basin**. 2nd ed. Los Angeles: Natural History Museum of Los Angeles County, 1993. 448p. $45.00; $27.95pa. ISBN 0938644297pa; 0938644327.
Covers 500 species, including 170 arachnids; color photos.

862. Hostetler, Mark. **That Gunk on Your Car: A Unique Guide to Insects of the United States**. Berkeley, CA: Ten Speed, 1997. 104p. $10.00pa. ISBN 089815961Xpa.
25 major groups of insects; color illustrations of the "splats" on windshields and the insects that make them; natural history. Also includes games and activities for travelers. Illustrated by Rebekah McClean.

863. Opler, Paul A. **A Field Guide to Western Butterflies**. 2nd ed. Boston: Houghton Mifflin, 1998. 528p. (Peterson Field Guide Series). ISBN 0395791529; 0395791510pa.
590 species; black-and-white and color photos; maps; checklist; plates separate from descriptions; extensive introduction. Covers area west of 100th meridian, including Hawaii and Alaska. Revised edition of *Field Guide to Western Butterflies* by James W. Tilden and Arthur Clayton Smith.

864. Powell, Jerry A., and Charles L. Hogue. **California Insects**. Berkeley, CA: University of California Press, 1979. 388p. (California Natural History Guides, No. 44). $16.95pa. ISBN 0520038061; 0520037820pa.
600 species; black-and-white illustrations and color photos. Illustrated by Charles L. Hogue.

865. Pyle, Robert Michael. **Watching Washington Butterflies: An Interpretive Guide to the State's 134 Species, Including Most of the Butterflies of Oregon, Idaho, and British Columbia**. Seattle, WA: Seattle Audubon Society, 1974. 109p. (Trailside Series). ISBN 0914516035.

  134 species; color photos; checklist.

866. Stoops, Erik D., and Jeffrey L. Martin. **Scorpions and Venomous Insects of the Southwest**. Phoenix, AZ: Golden West, 1995. 108p. $9.95pa. ISBN 0914846876pa.

  50 species; line drawings and color photos; includes first aid.

Fishes

Fish field guides are very popular, and a number have been created using plastic or laminated cards that can be taken to the beach or even in the water while snorkeling or scuba diving. Many field guides that cover game fishes also include angling tips or information on record sizes. Due to the economic importance of fishes, there are a number of semi-technical guides for the identification of fishes that are commercially harvested, many of them by the Food and Agriculture Organization of the United Nations (FAO).

Marine and freshwater fishes are included here, as are sharks and rays. As with the marine invertebrate guides, the fishes of southern Florida are often included with the fishes of the Caribbean islands and are thus found in the Central and South American section. Fishes can also be found in Chapter 2, "Flora and Fauna," and in Chapter 9, "Animals." Laminated guides are not covered here, but can be found at http://www.library.uiuc.edu/bix/fieldguides/main.htm.

NORTH AMERICA

❑ Boschung, Herbert T., David K. Caldwell, Melba C. Caldwell, Daniel W. Gotshall, and James D. Williams. **The Audubon Society Field Guide to North American Fishes, Whales, and Dolphins**. See entry 710.

867. Castro, Jose I. **The Sharks of North American Waters**. College Station: Texas A&M University Press, 1983. 180p. $19.95pa. ISBN 0890961409; 0890961433pa.

108 species; keys; line drawings and color photos, including drawings of shark teeth.

868. Dalrymple, Byron W. **Complete Guide to Game Fish: A Field Book of Fresh- and Saltwater Species**. 2nd ed. New York: Van Nostrand Reinhold, 1981. 506p. (Outdoor Life Books). ISBN 0442219784.

300 species; black-and-white illustrations; background and angling information. Originally published as *Sportsman's Guide to Game Fish*.

869. Dingerkus, Guido. **The Shark Watchers' Guide**. New York: Simon and Schuster, 1985. 144p. ISBN 0671550381pa.

38 species; black-and-white illustrations; background information and natural history.

870. Eddy, Samuel, and James C. Underhill. **How to Know the Freshwater Fishes**. 3rd ed. Dubuque, IA: W. C. Brown, 1978. (Pictured Key Nature Series). $25.80pa. ISBN 0697047512; 0697047504pa.

600 species; keys; line drawings.

871. Elman, Robert. **The Fisherman's Field Guide to the Freshwater and Saltwater Gamefish of North America**. New York: Knopf, 1977. 527p. ISBN 0394413997.

80 species of fish; black-and-white and some color photos; extensive natural history and angling tactics; includes record fish (as of 1977). Definitely for the angler.

872. Fautin, Daphne G., and Gerald R. Allen. **Field Guide to Anemone Fishes and Their Host Sea Anemones**. Perth: Western Australia Museum, 1992. 160p. $30.00. ISBN 0730952169.

10 anemones and 29 anemone fishes; visual key; color photos and illustrations; maps. Natural history and aquarium care. At the time of writing, the full text and illustrations from this field guide were available on the Worldwide Web, at http://www.keil.ukans.edu/ebooks/intro.html. Winner of 1993 Whitley Certificate for Best Field Guide.

873. Fichter, George S., and Phil Francis. **A Guide to Fresh and Salt-Water Fishing**. Rev. ed. New York: Golden, 1987. 160p. (Golden Guide). $4.50pa. ISBN 0307240509pa.

140 species; color illustrations, some color photos; about one-third of the volume consists of identification, the remainder has guides to bait and tackle, and tips on fishing.

874. Filisky, Michael. **Peterson First Guide to Fishes of North America**. Boston: Houghton Mifflin, 1989. 128p. $4.95pa. ISBN 0395502195pa.

    175 species; color illustrations.

875. McClane, A. J. **McClane's Field Guide to Freshwater Fishes of North America**. New York: Holt, Rinehart and Winston, 1978. 212p. ISBN 003015426X.

    224 species; color illustrations.

876. McClane, A. J. **McClane's Field Guide to Saltwater Fishes of North America: A Project of the Gamefish Research Association**. New York: Holt, Rinehart and Winston, 1978. 283p. ISBN 0030211212.

    400 species; color and black-and-white illustrations.

877. Page, Lawrence M., and Brooks M. Burr. **A Field Guide to Freshwater Fishes: North America North of Mexico**. Boston: Houghton Mifflin, 1991. 432p. (Peterson Field Guide Series, No. 42). $18.00pa. ISBN 0395910919pa.

    790 species; color and black-and-white illustrations; maps. Descriptions include range, remarks, and similar species.

878. Robison, Henry W. **Freshwater Fish**. New York: Smithmark, 1992. 192p. (American Nature Guides). $9.98. ISBN 0831769688.

    Not seen; see series entry 2.

879. Romashko, Sandra. **Handbook of Saltwater Fishes: Atlantic, Gulf, and Caribbean**. Miami, FL: Windward, 1992. 64p. (Sunshine Series). $3.95pa. ISBN 0893170402pa.

    110 species; black-and-white illustrations; pamphlet. Includes notes on edibility and angling tips.

880. Thompson, Peter. **Thompson's Guide to Freshwater Fishes: How to Identify the Common Freshwater Fishes of North America, How to Keep Them in a Home Aquarium**. Boston: Houghton Mifflin, 1985. 205p. ISBN 0395318386; 0395378036pa.

    113 species; color illustrations separate from text; maps; numerous common names; natural history. Does not include angling tips, but does cover keeping native fish as pets.

881. Zim, Herbert Spencer, and Hurst H. Shoemaker. **Fishes: A Guide to Fresh- and Salt-Water Species: 278 Fishes in Full Color**. New York: Golden, 1987. 160p. (Golden Guide). $5.95pa. ISBN 0307640590; 0307240592pa.

    278 species worldwide; color illustrations.

Eastern North America

882. Chilton, Earl W., II. **Freshwater Fishes of Texas**. Austin: Texas Parks and Wildlife, 1997. 98p. $12.95pa. ISBN 1885696221pa.
       46 species; color photos; descriptions include distribution, natural history, and angling tips.

883. Coad, Brian W. **Guide to the Marine Sport Fishes of Atlantic Canada and New England**. Toronto: University of Toronto Press, 1992. 307p. $50.00; $19.95pa. ISBN 0802058752; 0802067980pa.
       70 species of bait and sport fishes; black-and-white illustrations; maps; natural history.

884. Goodson, Gar. **Fishes of the Atlantic Coast: Canada to Brazil, Including the Gulf of Mexico, Florida, Bermuda, the Bahamas, and the Caribbean: 408 Fishes in Full Color**. Stanford, CA: Stanford University Press, 1985. 204p. $8.95pa. ISBN 0804712689pa.
       408 species; color illustrations; natural history; common names in English and Spanish. Revision of *The Many-Splendored Fishes of the Atlantic Coast*. Illustrated by Phillip J. Weisgerber.

885. Hoese, H. Dickson, and Richard H. Moore. **Fishes of the Gulf of Mexico, Texas, Louisiana, and Adjacent Waters**. 2nd ed. College Station: Texas A&M University Press, 1998. 422p. (W. L. Moody, Jr., Natural History Series, No. 22). ISBN 0890967377; 0890967679pa.
       540 species; keys; black-and-white illustrations and color photos.

886. Humann, Paul. **Reef Fish Identification: Florida, Caribbean, Bahamas**. 2nd ed. Jacksonville, FL: New World, 1994. 396p. $39.95pa. ISBN 1878348078pa; 1878348140.
       494 species; color photos; checklist; waterproof covers. Spiralbound.

887. Humann, Paul. **Reef Fish Identification: Florida, Caribbean, Bahamas: The Ultimate Fish Learning Tool**. Version 1.2C. Jacksonville, FL: New World, 1997. $39.95 (Windows).
       Based on the book *Reef Fish Identification: Florida, Caribbean, Bahamas* (see entry 886). Includes text and images from the book, a life list, plus a number of other features dealing with snorkeling and fish identification.

        ❑ Humann, Paul, and Ned DeLoach. **Snorkeling Guide to Marine Life: Florida, Caribbean, Bahamas**. See entry 719.

❑   Kaplan, Eugene H. **A Field Guide to Coral Reefs of the Caribbean and Florida: A Guide to the Common Invertebrates and Fishes of Bermuda, the Bahamas, Southern Florida, the West Indies, and the Caribbean Coast of Central and South America**. See entry 721.

888.   King, Everett Louis, and Thomas A. Edsall. **Illustrated Field Guide for the Classification of Sea Lamprey Attack Marks on Great Lakes Lake Trout**. Ann Arbor, MI: Great Lakes Fishery Commission, 1979. 41p. (Special Publication of the Great Lakes Fishery Commission, No. 79–1).

A small, spiralbound publication showing the two major types of sea lamprey attack marks and the four stages of healing for each type; color photos. One of the more specialized field guides.

889.   Laerm, Joshua, and B. J. Freeman. **Fishes of the Okefenokee Swamp**. Athens, GA: University of Georgia Press, 1986. 118p. $15.00; $7.50pa. ISBN 082030820X; 0820308412pa.

36 species; line drawings; lengthy entries, including angling information.

890.   Manooch, Charles S. **Fisherman's Guide: Fishes of the Southeastern United States**. Raleigh, NC: North Carolina State Museum of Natural History, 1984. 362p. ISBN 0917134079.

150 species; color illustrations; maps; fishing and eating tips.

❑   Martinez, Andrew J. **Marine Life of the North Atlantic: Canada to New England**. See entry 741.

891.   Phillips, Gary L., William D. Schmid, and James C. Underhill. **Fishes of the Minnesota Region**. Minneapolis: University of Minnesota Press, 1982. 248p. $14.95pa. ISBN 0816609799; 0816609829pa.

148 species; color photos.

892.   Roberts, Mervin F. **The Tidemarsh Guide to Fishes**. Old Saybrook, CT: Saybrook, 1985. 373p. $10.95pa. ISBN 0961504706pa.

100 species; line drawings; natural history. Covers area from Nova Scotia to Florida.

893.   Robins, C. Richard. **Saltwater Fish**. New York: Smithmark, 1992. 192p. (American Nature Guides). ISBN 0831769718.

300 species; color illustrations; spiralbound.

894. Robins, C. Richard, and G. Carleton Ray. **A Field Guide to Atlantic Coast Fishes of North America**. Boston: Houghton Mifflin, 1986. 354p. (Peterson Field Guide Series, No. 32). $20.95; $16.95pa. ISBN 0395318521; 0395391989pa.

    Comprehensive (1,000 species); black-and-white illustrations, some color.

895. Rohde, Fred C., et al. **Freshwater Fishes of the Carolinas, Virginia, Maryland, and Delaware**. Chapel Hill: University of North Carolina Press, 1994. 222p. $34.95; $16.95pa. ISBN 0807821306; 0807845795pa.

    260 species; color photos; maps; natural history.

896. Smith, C. Lavett. **National Audubon Society Field Guide to Tropical Marine Fishes of the Caribbean, the Gulf of Mexico, Florida, the Bahamas, and Bermuda**. New York: Knopf, 1997. 720p. $20.50 (flexi). ISBN 067944601X (flexi).

    400 species covered in detail, another 800 mentioned; visual keys; color photos and black-and-white illustrations; maps. Includes identification, habitat, range, notes, and similar species.

897. Stokes, F. Joseph. **Divers and Snorkelers Guide to the Fishes and Sea Life of the Caribbean, Florida, Bahamas, and Bermuda**. Rev. ed. Philadelphia: Philadelphia Academy of Natural Sciences, 1984. 160p. $14.95pa. ISBN 0910006466; 069001919Xpa.

    460 species; color illustrations; quick reference guide to fishes with obvious characteristics such as color or form. Revision of *Handguide to the Coral Reef Fishes of the Caribbean*, 1980. Illustrated by Charlotte C. Stokes.

898. Walls, Jerry G. **Fishes of the Northern Gulf of Mexico**. Hong Kong: T. F. H., 1975. 432p. ISBN 0876664451.

    502 species; line drawings and color photos; large for a field guide but includes separate waterproof plastic sheet containing color illustrations of 256 common species.

899. Werner, Robert G. **Freshwater Fishes of New York State: A Field Guide**. Syracuse, NY: Syracuse University Press, 1980. 186p. (A York State Book). ISBN 0815622228.

    Comprehensive, with keys to all species and 68 common species illustrated with line drawings.

Western North America

900.   Armstrong, Robert H. **Alaska's Fish: A Guide to Selected Species**. Anchorage: Alaska Northwest, 1996. 94p. (Alaska Pocket Guide). $12.95pa. ISBN 0882404725pa.

19 freshwater and 15 saltwater species; color photos and line drawings; includes description, habits, range. Pocket sized.

901.   California Department of Fish and Game. **Marine Sportfish Identification, California**. Sacramento, Ca: Dept. of Fish and Game, 1987. 164p.

81 species; color illustrations and line drawings; natural history.

902.   Carpenter, Russell B. and Blyth C. Carpenter. **Fish Watching in Hawaii**. San Mateo, CA: Natural World, 1981. 120p. ISBN 0939560003pa; 0939560011.

85 species; color illustrations; about one-half of the book consists of natural history with color photos.

903.   Eschmeyer, William N., and Earl S. Herald. **A Field Guide to Pacific Coast Fishes of North America: From the Gulf of Alaska to Baja California**. Boston: Houghton Mifflin, 1983. 336p. (Peterson Field Guide Series, No. 28). $17.95pa. ISBN 0395268737; 0395331889pa.

500 species; color and black-and-white illustrations.

904.   Fitch, John E., and Robert J. Lavenberg. **Tidepool and Nearshore Fishes of California**. Berkeley, CA: University of California Press, 1975. 156p. (California Natural History Guides, No. 38). $7.95pa. ISBN 0520028449; 0520028457pa.

80 species; keys to family or genus; line drawings.

905.   Fitch, John E., and Robert J. Lavenberg. **Marine Food and Game Fishes of California**. Berkeley, CA: University of California Press, 1971. 179p. (California Natural History Guides, No. 28). $10.95pa. ISBN 0520018311pa.

200 species; black-and-white illustrations and color photos; natural history.

906.   Goodson, Gar. **Fishes of the Pacific Coast: Alaska to Peru, Including the Gulf of California and the Galapagos Islands**. Stanford, CA: Stanford University Press, 1988. 267p. $10.95pa. ISBN 0804713855pa.

433 species illustrated, another 46 only described; color illustrations.

907.   Goodson, Gar. **The Many-Splendored Fishes of Hawaii: 170 Fishes in Full Color**. Palos Verdes Estates, CA: Marquest Colorguide, 1973. 91p. $7.95pa. ISBN 0804712700pa.

166 species; color illustrations; common names in English and Hawaiian; pocket sized.

❏ Gotshall, Daniel. **Marine Animals of Baja California: A Guide to the Common Fish and Invertebrates**. 2nd ed. See entry 746.

908.   Gotshall, Daniel. **Pacific Coast Inshore Fishes**. 2nd, Completely rev. ed. Los Osos, CA: Sea Challengers, 1981. 96p. $18.95pa. ISBN 0930118065pa.

193 species; visual key; color photos. Revised edition of *Fishwatchers' Guide to the Inshore Fishes of the Pacific Coast*.

909.   Hoover, John P. **Hawaii's Fishes: A Guide for Snorkelers, Divers, and Aquarists**. Honolulu: Mutual, 1993. 178p. $18.95pa. ISBN 1566470013pa.

230 species; color photos; common names in English and Hawaiian; includes descriptions of the best snorkeling and diving sites. Arranged by family; each family has a one to two page introduction, followed by attractive color photos of selected species with brief descriptions and location information. Also includes extensive bibliography and a list of invalid scientific names.

910.   Humann, Paul. **Coastal Fish Identification: California to Alaska**. Jacksonville, FL: New World, 1996. 205p. $32.95 (comb binding). ISBN 1878348124 (comb binding); 1878348175.

250 species; color photos; waterproof covers.

❏ Kessler, Doyne W. **Alaska's Saltwater Fishes and Other Sea Life: A Field Guide**. See entry 750.

911.   Lamb, Andy, and Phil Edgell. **Coastal Fishes of the Pacific Northwest**. Madeira Park, BC, Canada: Harbor, 1986. 224p. $23.95pa. ISBN 0920080758pa.

175 species; color photos and black-and-white illustrations; natural history including fishing tips, use, and distribution. Multiple common names.

912.   Love, Robin Milton. **Probably More Than You Want to Know About the Fishes of the Pacific Coast**. Santa Barbara, CA: Really Big, 1991. 215p. $19.95. ISBN 0962872547.

200 species; black-and-white illustrations; extensive natural history; written in a very casual style.

913. Mahaney, Casey. **Hawaiian Reef Fishes: The Identification Book**. Kailua-Kona, HI: Blue Kirio, 1993. 125p. $19.95pa. ISBN 0963625209pa.

160 species; color photos; descriptions include habitat and behavior.

914. McGinnis, Samuel M. **Freshwater Fishes of California**. Berkeley, CA: University of California Press, 1984. 316p. (California Natural History Guides, No. 49). $15.95pa. ISBN 0520048814; 0520048911pa.

125 species; pictorial key; black-and-white illustrations; checklist; natural history; and angling information.

915. Pijoan, Michel. **Game Fish of the Rocky Mountains: A Guide to Identification and Habitat**. Flagstaff, AZ: Northland, 1985. 68p. ISBN 0873583728pa.

36 species; color illustrations.

916. Randall, John E. **Guide to Hawaiian Reef Fishes**. Newtown Square, PA: Harrowood, 1985. 74p. $18.95 (plastic). ISBN 0915180073; 0915180294pa; 0915180022 (plastic).

177 species; color photos separate from text. Waterproof version available as *Underwater Guide to Hawaiian Reef Fishes*, containing photos only.

917. Randall, John E. **Shore Fishes of Hawaii**. Vida, OR: Natural World, 1996. $14.75pa. ISBN 0939560216pa.

340 species; color photos; checklist; common names in English and Hawaiian.

918. Russo, Ron. **Pacific Coast Fish: A Guide to Marine Fish of the Pacific Coast of North America**. Berkeley, CA: Nature Study Guild, 1990. 107p. $4.75pa. ISBN 0912550198pa.

145 species; visual key to families; black-and-white illustrations; pocket sized.

919. Somerton, David, and Craig Murray. **Field Guide to the Fish of Puget Sound and the Northwest Coast**. Seattle, WA: Distributed by the University of Washington Press, 1976. 70p. (A Washington Sea Grant Publication). ISBN 0295954973.

200 species; black-and-white illustrations; waterproof plastic.

920. Taylor, Leighton R. **Sharks of Hawaii: Their Biology and Cultural Significance**. Honolulu: University of Hawaii Press, 1993. 126p. $23.95. ISBN 0824815629.

    37 species; line drawings and color photos; half of the book is background information; anthropology.

921. Witte, Astrid. **Hawaii: Reef Fish and Critter I.D.: Snorkel Skills and Professional Tips**. Kailua-Kona, HI: Blue Kirio, 1996. 75p. (Reef Watcher's). $9.95pa. ISBN 0965348601.

    Not seen; see series entry 88. In English and Japanese. Title in Japanese: *Hawai: Sunokeringu no Gijutsu to Puro no Adobaisu.*

Chapter 13

Reptiles and Amphibians

Reptiles and amphibians (known to connoisseurs as herps, from the Greek *herpetos*) are a group of animals that often produce an exaggerated fear in many people, despite the inoffensive nature of most species. There are relatively few species of reptiles and amphibians, especially in temperate areas, so most herp field guides have space for natural history and do their best to ease some of this unreasonable concern. A number of the guides listed in this chapter have been published by university presses, and are more technical in nature than most field guides.

Since there are several venomous species, many herp field guides include information on which species are dangerous and how to treat bites. The first aid information found in older guides should be treated with caution, since treatment recommendations have changed over the years. Herp field guides often have information on keeping reptiles and amphibians as pets, which should also be treated with caution. World populations of many amphibians in particular are threatened, if not endangered, so only the most common species should be taken from the wild.

Other guides that include reptile and amphibians are found in Chapter 2, "Flora and Fauna," and Chapter 9, "Animals."

NORTH AMERICA

922.   Ballinger, Royce E., and John D. Lynch. **How to Know the Amphibians and Reptiles**. Dubuque, IA: W. C. Brown, 1983. 229p. (Pictured Key Nature Series). $19.45pa. ISBN 0697047865pa.

Comprehensive; key throughout, including separate key to larval amphibians; black–and–white illustrations; maps.

923. Behler, John. **The Audubon Society Field Guide to North American Reptiles and Amphibians**. New York: Knopf, 1979. 719p. (Audubon Society Field Guide Series). $19.00 (flexi). ISBN 0394508246 (flexi).

    400 species; maps; color photos.

924. Behler, John, ed. **Familiar Reptiles and Amphibians**. New York: Knopf, 1988. 192p. (Audubon Society Pocket Guides). $9.00pa. ISBN 0394757939pa.

    80 species illustrated, another 80 similar species mentioned; color photos; maps.

925. Capula, Massimo. **Simon and Schuster's Guide to Reptiles and Amphibians of the World**. New York: Simon and Schuster, 1989. 256p. $14.00pa. ISBN 0671691368; 0671690981pa.

    200 species worldwide; color photos.

926. Conant, Roger, Robert Stebbins, and Joseph T. Collins. **Peterson First Guide to Reptiles and Amphibians**. Boston: Houghton Mifflin, 1992. 128p. $5.95pa. ISBN 0395622328pa.

    250 species; color illustrations.

927. Cook, Francis R. **Introduction to Canadian Amphibians and Reptiles**. Ottawa: National Museum of Natural Sciences, National Museums of Canada, 1984. 200p. ISBN 0660107554.

    90 species; black-and-white illustrations; maps; natural history and a chapter on caring for amphibian and reptile pets. Illustrated by Charles Douglas.

928. Forey, Pamela, and Cecilia Fitzsimons. **An Instant Guide to Reptiles and Amphibians: The Most Familiar Species of North American Reptiles and Amphibians Described and Illustrated in Color**. New York: Bonanza, 1987. 124p. ISBN 0517618001.

    140 species; color illustrations; checklist.

929. Froom, Barbara. **Amphibians of Canada**. Toronto: McClelland and Stewart, 1982. 120p. ISBN 0771032072.

    22 species; color photos; natural history. Includes information on keeping amphibians as pets.

    ❏ Reader's Digest Editors. **North American Wildlife: Mammals, Reptiles, and Amphibians**. See entry 725.

930. Scott, Chris. **Snake Lovers' Lifelist and Journal**. Austin: University of Texas Press, 1996. 246p. $19.95. ISBN 0292776985.

All known species of snakes in the U.S. and Canada, plus subspecies and hybrids (438 types, total); color photos and line drawings. Oversized; primarily a journal with space for notes on sitings of each species, but includes identification.

931. Smith, Hobart Muir. **Amphibians of North America: A Guide to Field Identification**. New York: Golden, 1978. 160p. (Golden Field Guide Series). ISBN 0307136620.

183 species; keys; color illustrations; maps. Illustrated by Sy Barlowe.

932. Smith, Hobart Muir, and Edmund D. Brodie, Jr. **Reptiles of North America: A Guide to Field Identification**. New York: Golden, 1982. 240p. (Golden Field Guide Series). $11.95pa. ISBN 0307470091; 0307136663pa.

278 species; brief visual keys; color illustrations; maps; background information. Illustrated by David M. Dennis and Sy Barlowe.

933. Tyning, Thomas F. **A Guide to Amphibians and Reptiles**. Boston: Little, Brown, 1990. 400p. (Stokes Nature Guide). $19.95; $14.95pa. ISBN 0316817198; 0316817139pa.

32 species; black-and-white illustrations; maps; extensive natural history. Illustrated by Andrew Finch Magee.

934. U.S. Department of the Navy, Bureau of Medicine and Surgery. **Poisonous Snakes of the World**. New York: Dover, 1991. 203p. $14.95pa. ISBN 048626629Xpa.

Covers all genera worldwide, with emphasis on the most common or dangerous species; keys; black-and-white photos and illustrations, some color illustrations; oversized. Arranged by region.

935. Zim, Herbert Spencer, and Hobart M. Smith. **Reptiles and Amphibians: 212 Species in Full Color**. New York: Golden, 1991. 160p. (Golden Guide). $5.95pa. ISBN 0307640574; 0307240576pa.

Color illustrations; maps. Illustrated by James G. Irving.

Eastern North America

936. Conant, Roger and Joseph T. Collins. **A Field Guide to Reptiles and Amphibians: Eastern and Central North America**. 3rd ed., exp. Boston: Houghton Mifflin, 1998. 616p. (Peterson Field Guide Series, No. 12). $20.00pa. ISBN 0395904528pa.

574 species; color illustrations, some black-and-white photos and line drawings; maps; covers area west to the Dakotas. Descriptions include ranges, similar species, and subspecies. Also has information on capturing and keeping herps as pets. First edition published as *A Field Guide to Reptiles and Amphibians of the United States and Canada East of the 100th Meridian*. Illustrated by Isabelle Hunt Conant.

937.   Elliott, Lang. **Calls of Frogs and Toads**. Ithaca, NY: NatureSound Studio, 1992. Sound cassette or compact disk and booklet. $13.95 (cassette); $16.95 (CD). ISBN 1878194240 (cassette); 1878194259 (CD).

Covers calls of 40 species of eastern and central North American frogs and toads.

938.   Epple, Anne Orth. **The Amphibians of New England**. Camden, ME: Down East, 1983. 138p. ISBN 0892721596pa.

22 species; black-and-white illustrations; natural history. Illustrated by Patrice Rossi.

939.   Garrett, Judith M., and David G. Barker. **A Field Guide to Reptiles and Amphibians of Texas**. Austin, TX: Texas Monthly, 1987. 225p. (Texas Monthly Field Guide Series). $21.95; $14.95pa. ISBN 0877190682; 0877190917pa.

167 species; color photos; maps; natural history.

940.   Haast, William E., and Robert Anderson. **Complete Guide to Snakes of Florida**. Miami, FL: Phoenix, 1981. 139p. $9.95pa. ISBN 09408100Xpa.

63 species; color photos; natural history.

941.   Harding, James H. **Amphibians and Reptiles of the Great Lakes Region**. Ann Arbor: University of Michigan Press, 1997. 378p. (Great Lakes Environment). $42.50; $27.50pa. ISBN 0472096281; 0472066285pa.

30 species of amphibians, 40 of reptiles; color photos; maps; extensive introduction. Species accounts include description, confusing species, distribution and status, habitat and ecology, reproduction, and conservation.

942.   Harding, James H., and J. Alan Holman. **Michigan Turtles and Lizards: A Field Guide and Pocket Reference**. 2nd ed. East Lansing, MI: Michigan State University, Cooperative Extension Service, 1997. 94p. (Extension Bulletin, E–2234). $8.95pa. ISBN 1565250117pa.

10 turtles and 2 lizards; key; color photos; checklist; maps; natural history.

943. Harding, James H., and J. Alan Holman. **Michigan Frogs, Toads, and Salamanders: A Field Guide and Pocket Reference**. East Lansing, MI: Michigan State University, Cooperative Extension Service, 1992. 144p. (Extension Bulletin, E–2350). $15.95; $11.95pa. ISBN 1565250028pa; 1565250036.

23 species; color photos; maps; checklist; natural history.

944. Holman, J. Alan. **Michigan Snakes: A Field Guide and Pocket Reference**. East Lansing, MI: Michigan State University Extension, 1989. 72p. (Extension Bulletin, E–2000). $6.95pa. ISBN 1565250044pa.

17 species; simple key; color photos; maps; checklist; natural history.

945. Martof, Bernard, et al. **Amphibians and Reptiles of the Carolinas and Virginia**. Chapel Hill: University of North Carolina Press, 1980. 264p. $16.95pa. ISBN 0807813893; 0807842524pa.

200 species; color photos; maps.

946. Price, Andrew H. **Poisonous Snakes of Texas**. Rev. ed. Austin: Texas Parks and Wildlife, 1998. 112p. $13.95pa. ISBN 1885696221pa.

16 species or subspecies; color photos and black–and–white illustrations; maps; natural history; first aid. Revision of John F. Werler's book by the same name.

947. Tennant, Alan. **A Field Guide to Snakes of Florida**. Houston, TX: Gulf, 1997. 272p. (Gulf's Field Guide Series). $21.95pa. ISBN 087719291Xpa.

70 species; color photos and line drawings; maps; extensive natural history.

948. Tennant, Alan. **Gulf Field Guide to Texas Snakes**. 2nd ed. Austin, TX: Texas Monthly, 1998. 290p. $21.95pa. ISBN 0877192774pa.

105 species; keys; color photos separate from text, black–and–white illustrations; maps; natural history. Includes poster comparing venomous and non-venomous snakes.

Western North America

949. Brown, Philip R. **A Field Guide to Snakes of California**. Houston, TX: Gulf, 1997. 215p. (Gulf's Field Guide Series). $21.95pa. ISBN 0877193088pa.

    38 species; key; color photos and black-and-white illustrations; maps; checklist. Common names in English and Spanish.

950. Brown, Herbert A., Robert M. Storm, and William P. Leonard. **Reptiles of Washington and Oregon**. Seattle, WA: Seattle Audubon Society, 1995. 176p. (Trailside Series). ISBN 0914516124.

    180 species; color and black-and-white illustrations.

951. Corkran, Charlotte C., and Chris Thoms. **Amphibians of Oregon, Washington and British Columbia: A Field Identification Guide**. Redmond, WA: Lone Pine, 1996. 175p. $16.95pa. ISBN 1551050730pa.

    36 species; keys to eggs, hatchlings, larvae, tadpoles, and adult stages; color photos; maps; includes extensive background information on amphibian biology, locating, photographing, and surveying amphibians, plus descriptions of habitats, eggs, hatchlings, larvae, tadpoles, and adult stages. A good, comprehensive guide.

952. Koch, Edward D. **Amphibians and Reptiles of Yellowstone and Grand Teton National Parks**. Salt Lake City, UT: University of Utah Press, 1995. 188p. $12.95pa. ISBN 0874804728pa.

    6 amphibian and 6 reptile species; simple key; color photos; maps; extensive natural history.

953. Leonard, William P., et al. **Amphibians of Washington and Oregon**. Seattle, WA: Seattle Audubon Society, 1993. 168p. (Trailside Series). $14.95pa. ISBN 0914516108pa.

    33 species; color photos; natural history.

954. Lowe, Charles H., Cecil R. Schwalbe, and Terry B. Johnson. **The Venomous Reptiles of Arizona**. Phoenix, AZ: Arizona Game and Fish Dept., 1986. 115p. ISBN 0917563034pa.

    19 species; color photos; natural history; extensive background and first aid.

955. McKeown, Sean. **A Field Guide to Reptiles and Amphibians in the Hawaiian Islands**. Los Osos, CA: Diamond Head, 1996. 172p. $15.95pa. ISBN 0965073106pa.

    33 species; keys; color photos; maps; common names in English and Hawaiian. Descriptions include reproduction, distribution, general notes, and care in captivity.

956. Nussbaum, Ronald A., Edmund D. Brodie, Jr., and Robert M. Storm. **Amphibians and Reptiles of the Pacific Northwest**. Moscow, ID: University Press of Idaho, 1983. 332p. (Northwest Naturalist Books). $24.95pa. ISBN 0893010863.

All 60 species; key; black-and-white photos; maps.

957. Russell, Anthony P., and Aaron M. Bauer. **The Amphibians and Reptiles of Alberta: A Field Guide and Primer of Boreal Herpetology**. Calgary, Alta.: University of Calgary Press; Edmonton, Alta.: University of Alberta Press, 1993. 264p. $24.95pa; $29.95. ISBN 1895176468 (U. of C. pa); 088864261X (U. of A. hb); 0888642628 (U. of A., pa); 1895176204 (U. of C. hb).

18 species plus 4 possibly found in Alberta; keys to adult, larvae, and eggs; black-and-white photos, separate color photos; maps; checklist; natural history. Extensive background information.

958. Stebbins, Robert C. **A Field Guide to Western Reptiles and Amphibians: Field Marks of All Species in Western North America, Including Baja California**. 2nd ed. rev. Boston: Houghton Mifflin, 1985. 336p. (Peterson Field Guide Series, No. 16). $24.95; $18.00pa. ISBN 0395382548; 039538253Xpa.

244 species and 260 subspecies; black-and-white and color illustrations. Illustrated by the author.

959. Stoops, Erik D., and Annette Wright. **Snakes and Other Reptiles of the Southwest**. 2nd ed. Phoenix, AZ: Golden West, 1993. 102p. $9.95pa. ISBN 0914846795pa.

65 species; black-and-white and color photos.

960. Williamson, Michael A., Paul W. Hyder, and John S. Applegarth. **Snakes, Lizards, Turtles, Frogs, Toads, and Salamanders of New Mexico**. Santa Fe, NM: Sunstone, 1994. 176p. $19.95pa. ISBN 0865342334.

123 species; keys; color photos; maps; natural history.

Birds

There are indisputably more field guides to birds than there are for any other group of plants or animals, with only guides to wildflowers even coming close to the popularity of bird guides. Since there are so many field guides, authors and publishers have had to try to find a niche, so there are a larger variety of types of bird field guides than there are for the other guides. Books range from beginner's guides to extremely advanced guides intended for the use of experts wanting to identify what are sometimes known as LBJs (Little Brown Jobs). See the Introduction for a discussion of types of bird field guides.

The illustrations in bird field guides may be either photographs or paintings. Many experts firmly believe that a good illustration is better than the best photograph, though this is not a universal opinion. Many bird field guide illustrators are well known as artists, including Roger Tory Peterson, Arthur Singer, and Lars Jonsson. Louis Agassiz Fuertes, considered by some to be a better ornithological artist than John James Audubon, also illustrated field guides.

Bird finding guides, checklists, and other allied birding tools are excluded from consideration here. Of the many multimedia birder's aids such as bird song recordings, audiovisual recordings, and computer CD-ROMs, only the most common guides are annotated, including guides that are associated with the major field guide series. Most of these multimedia resources are not intended to be used in the field.

See also Chapter 2, "Flora and Fauna," and Chapter 9, "Animals," for more guides. A few bird footprints are often included in the animal track guides in Chapter 9 and Chapter 15, "Mammals."

NORTH AMERICA

961. Austin, Oliver L. **Families of Birds**. New, rev. ed. New York: Golden, 1985. 200p. (Golden Field Guide Series). ISBN 0307136698pa.

Covers all 208 families of birds worldwide (including 36 fossil families); color illustrations. Includes number of species, size of range, world distribution, sexual differences, and nesting habits for each family. Also published in Spanish by Editorial Trillas, 1994. Illustrated by Arthur B. Singer.

962. Baicich, Paul J., and Colin J. O. Harrison. **A Field Guide to the Nests, Eggs, and Nestlings of North American Birds**. 2nd ed. San Diego: Academic, 1997. 350p. $22.95pa. ISBN 0120728311pa.

597 species; keys; black-and-white and color illustrations; natural history. First edition by Harrison.

963. Bologna, Gianfranco. **Simon and Schuster's Guide to Birds of the World**. New York: Simon and Schuster, 1981. 511p. (A Fireside Book). $14.95pa. ISBN 0671422340; 0671422359pa.

424 species; color photos; arranged alphabetically by species; extensive introductory material. Translation of *Uccelli*.

964. Boyer, Trevor, and John Gooders. **Ducks**. New York: Gallery, 1990. 144p. (American Nature Guides). ISBN 0831769599.

Not seen; see series entry 2.

965. Burton, Philip John Kennedy. **Spotter's Guide to Birds of North America**. Usborne, 1991. 64p. (Usborne Spotter's Guides). $4.95pa. ISBN 0746011458pa.

170 species; color illustrations; checklist. Also published in the U.S. by Mayflower, 1979. Illustrated by Trevor Boyer and Tim Hayward.

966. Burton, Robert. **Concise Birdfeeder Handbook**. 1st American ed. New York: DK, 1997. 128 p. $8.95pa. ISBN 0789414651pa.

80 species; color photos; maps; information on attracting and feeding birds; the "field" for this handbook is the kitchen window. An abridged, pocket-sized version of the author's *National Audubon Society North American Birdfeeder Handbook*. Cover title: *National Audubon Society Concise Birdfeeder Handbook*.

967. Byers, Clive, Jon Curson, and Urban Olsson. **Sparrows and Buntings: A Guide to the Sparrows and Buntings of North America and the World**. Boston: Houghton Mifflin, 1995. 334p. $40.00. ISBN 0395738733.

86 species; color and black-and-white illustrations; maps. Extensive descriptions of males, females, and juveniles; an advanced guide. Illustrated by Clive Byers.

968. Chapman, Frank Michler. **Color Key to North American Birds**. Rev. ed. New York: D. Appleton and Co., 1912. 356p.

800 species; black-and-white illustrations; arranged chiefly by family but also by color (i.e., perching birds marked with red or chiefly dull colored). One of the earliest identification guides not intended for use with a bird in the hand (i.e., dead). Illustrated by Chester A. Reed, who also created his own field guides.

969. Clark, William S. **A Field Guide to Hawks of North America**. Boston: Houghton Mifflin, 1987. 198p. (Peterson Field Guide Series, No. 35). $24.95; $16.95pa. ISBN 0395360013; 0395441129pa.

39 species; color plates and black-and-white photographs of hawks in flight; natural history. Illustrated by Brian K. Wheeler.

970. Curson, Jon, David Quinn, and David Beadle. **Warblers of the Americas: An Identification Guide**. Boston: Houghton Mifflin, 1994. 252p. $40.00. ISBN 0395709989.

All 116 species of North, Central, and South America; color illustrations, maps, and brief identification information separate from main species account text. Originally published in the U.K. as part of the Christopher Helm Identification Guide series with the title *New World Warblers*.

971. Davis, L. Irby. **A Field Guide to the Birds of Mexico and Central America**. Austin: University of Texas Press, 1972. 282p. (John Fielding and Lois Lasater Maher Series, No. 1). ISBN 0292707002; 0292707029pa.

Comprehensive listing, though only species not found in North American field guides are illustrated and described. Illustrated by I. F. Bennett.

972. DeGraaf, Richard M., et al. **Forest and Rangeland Birds of the United States: Natural History and Habitat Use**. Washington, DC: U.S. Dept. of Agriculture, 1991. 625p. (Agriculture Handbook, No. 688).

515 species; color illustrations; descriptions include range, habitat requirements, nest, food, and habits; designed for foresters and rangeland managers. Illustrations taken from the Golden Field Guide.

973. Dunn, Jon, and Kimball Garrett. **A Field Guide to Warblers of North America**. Boston: Houghton Mifflin, 1997. 656p. (Peterson Field Guide Series, No. 49). $28.00; $20.00pa. ISBN 0395389712; 0395783216pa.

   60 species; color photos and illustrations; maps. More similar to the taxonomic-based bird guides published by Princeton University Press and others, than the usual Peterson guide; has plates of Peterson-type color illustrations facing brief descriptions in the front, followed by extensive species accounts that include descriptions, similar species, voices, behaviors, habitats, distribution, status and conservation, subspecies, taxonomic relationship, and plumages. Illustrated by Tom Schultz and Cindy House.

974. Dunne, Pete, David Sibley, and Clay Sutton. **Hawks in Flight**. Boston: Houghton Mifflin, 1988. 254p. $13.00pa. ISBN 0395423880pa.

   23 species; black-and-white illustrations and photos; based on overall impressions ("jizz"), not field marks. Advanced guide. Illustrated by David Sibley.

975. Edwards, Ernest Preston. **A Field Guide to the Birds of Mexico and Adjacent Areas: Belize, Guatemala and El Salvador**. 3rd ed. Austin, TX: University of Texas Press, 1998. 288p. $35.00; $17.95pa. ISBN 0292720920; 0292720912pa.

   1,000 species described, 850 species illustrated; color illustrations separate from text; common names in English and Spanish. Descriptions include range, abundance, and habitat. Illustrated by Edward Murrell Butler and John P. O'Neill.

976. Ehrlich, Paul R., David S. Dobkin, and Daryl Wheye. **The Birder's Handbook: A Field Guide to the Natural History of North American Birds, Including All Species That Regularly Breed North of Mexico**. New York: Simon and Schuster, 1988. 785p. $18.00pa. ISBN 0671621335; 0671659898pa.

   All 646 species of birds that breed in North America; a few black-and-white illustrations of behavior, flight silhouettes, eggs (but not birds). This is not a true field guide because the emphasis is on natural history, not identification; the book is also too heavy to carry easily into the field. Each species' behavior, food, nesting, etc. is covered in one-half page; facing pages contain short essays on a number of topics. Illustrated by Shahid Naeem.

977. Elliott, Lang, and Marie Read. **Common Birds and Their Songs**. Boston: Houghton Mifflin, 1998. 128p. $20.00. ISBN 0395912385. Compact disk and booklet.

50 species; color photos; natural history. Includes 60-minute compact disk with field recordings of songs.

978. Elliott, Lang. **Know Your Bird Sounds: Volume 1: Yard, Garden and City Birds**. Ithaca, NY: NatureSound Studios, 1991. Sound cassette or compact disk and booklet. $12.95 (cassette); $16.95 (CD). ISBN 1878194011 (cassette); 1878194194 (CD).

35 species.

979. Elliott, Lang. **Know Your Bird Sounds: Volume 2: Birds of the Countryside**. Ithaca, NY: NatureSound Studios, 1991. Sound cassette or compact disk and booklet. $12.95 (cassette); $16.95 (CD). ISBN 187819402X (cassette); 1878194216 (CD).

35 species.

980. Farrand, John. **How to Identify Birds**. New York: McGraw-Hill, 1988. 311p. (An Audubon Handbook). $17.95 (flexi). ISBN 0070199752 (flexi).

Not a true field guide; advanced identification tips for birders. Color photos.

981. Farrand, John, ed. **The Audubon Society Master Guide to Birding**. New York: Knopf, 1983. 3 v. $16.95 (v. 1, flexi); $15.95 (v. 2, flexi); $16.95 (v. 3, flexi). ISBN 0394533828 (v. 1, flexi); 0394533844 (v. 2, flexi); 0394533836 (v. 3, flexi).

Contents: Volume 1, Loons to Sandpipers; volume 2, Gulls to Dippers; volume 3, Old World Warblers to Sparrows. All 855 species of North America; many color photos, illustrations; maps; too heavy to carry into the field, but useful for detailed information on difficult identifications.

982. Fichter, George S. **Birds of North America: An Audubon Society Beginner's Guide**. New York: Random House, 1982. 96p. ISBN 0394847717pa.

64 species; color illustrations; checklist; pocket sized. A beginner's guide with extensive introductory material. Illustrated by Arthur Singer.

983. Grant, P. J. **Gulls: A Guide to Identification**. 2nd ed. San Diego: Academic, 1986. 352p. $34.00; $19.95pa. ISBN 0856610445; 0122956400pa.

31 species; black–and–white photos and line drawings; maps. Includes aging, molts, etc. Also published in the U.K. by Poyser.

984.   Griggs, Jack. **American Bird Conservancy's Field Guide to All the Birds of North America: A Revolutionary System Based on Feeding Behaviors and Field-Recognizable Features**. New York: HarperCollins, 1997. 172p. $19.95 (flexi). ISBN 0062730282 (flexi).

Comprehensive; visual key; color illustrations; maps; index/checklist. The keying system is based on how and where the birds feed, so it is arranged by groups such as aerialists, swimmers, and shorebirds. There are two to five species per page. Also published in the U.K. as *Collins Pocket Guide: Birds of North America*. Sections of the guide will also be published separately (see entries 1061 and 1127); volumes for 1999 include *All the Birds of Prey, All the Water Birds: Atlantic and Gulf Coasts, All the Water Birds: Pacific Coast*, and *All the Water Birds: Freshwater*.

985.   Hansard, Peter, and Burton Silver. **What Bird Did That? A Driver's Guide to Some Common Birds of North America**. Berkeley, CA: TenSpeed, 1991. 64p. $8.95pa. ISBN 0898154278pa.

40 species of North American birds are identified based on their "splays" (droppings) collected from automotive windshields.

986.   Harrison, Peter. **A Field Guide to Seabirds of the World**. Lexington, MA: Stephen Greene, 1987. 317p. $24.95. ISBN 0828906106pa.

321 species; color photos; maps; section with illustrations first, followed by species accounts. Advanced guide.

987.   Harrison, Peter. **Seabirds, an Identification Guide**. Rev. ed. Boston: Houghton Mifflin, 1985. 448p. $29.95pa. ISBN 0395332532; 03956062915pa.

312 species; color illustrations, some black-and-white; maps; advanced guide. Illustrated by the author.

988.   Harrison, Peter. **Seabirds of the World: A Photographic Guide**. Princeton, NJ: Princeton University Press, 1996. 317p. $24.95pa. ISBN 069101551lpa.

320 species; color photos; maps; includes photos of most species in flight. Designed as a pocket-sized companion to the author's *Seabirds, an Identification Guide*. Originally published in the U.K. as part of the Christopher Helm Identification Guide series.

989.   Hayman, Peter, John Marchant, and Tony Prater. **Shorebirds: An Identification Guide to the Waders of the World**. Boston: Houghton Mifflin, 1986. 412p. $29.95pa. ISBN 0395379032; 0395602378pa.

214 species of Charadriiformes; color illustrations; advanced birding guide in two sections, the first with color plates and brief descriptions and the second with species accounts and detailed

descriptions. Originally published in the U.K. as part of the Christopher Helm Identification Guide series.

990. Headstrom, Richard. **A Complete Field Guide to Nests in the United States, Including Those of Birds, Mammals, Insects, Fishes, Reptiles, and Amphibians**. New York: I. Washburn, 1970. 451p.

   700 species, mostly birds; keys; few illustrations; arranged by taxonomic group and location (eastern or western North America). Based on the author's *Birds' Nests, a Field Guide*, published 1949, and *Birds' Nests of the West, a Field Guide*, published 1951.

991. Hines, Bob. **Ducks at a Distance: A Waterfowl Identification Guide**. Washington: Dept. of the Interior, Fish and Wildlife Service, 1978. 53p.

   32 species of ducks and geese; numerous color and black-and-white illustrations of ducks in flight, including male and female plumage, plus plumage variations, flock patterns, and relative size.

992. Hines, Bob, ed. **Fifty Birds of Town and City**. Washington: U.S. Dept. of the Interior, Fish and Wildlife Service, 1993. 50p.

   Color illustrations; natural history. Illustrated by Hines.

993. Howell, Steven N. G., and Sophie Webb. **A Guide to the Birds of Mexico and Northern Central America**. New York: Oxford University Press, 1994. 851p. $79.00; $39.95pa. ISBN 0198540132; 0198540124pa.

   Advanced guide/manual covering all 1,070 species found in Mexico, Guatemala, Belize, El Salvador, Honduras, and western Nicaragua; color drawings. Extensive descriptions, including vocalizations and habits. Illustrated by Sophie Webb.

994. Jaques, H. E. **How to Know the Land Birds: Pictured-Keys for Determining All of the Land Birds of the Entire United States and Southern Canada, with Maps Showing Their Geographic Distribution and Other Helpful Features**. Dubuque, IA: W. C. Brown, 1947. 196p. (Pictured Key Nature Series).

   Comprehensive; key; black-and-white illustrations; maps.

995. Jaques, H. E., and Roy Ollivier. **How to Know the Water Birds: Pictured-Keys for Determining the Water Birds of the United States and Canada, with Maps Showing Their Geographic Distribution, and with Other Helpful Features**. Dubuque, IA: W. C. Brown, 1960. 159p. (Pictured Key Nature Series).

   225 species; key; black-and-white illustrations; maps.

996. Jaramillo, Alvaro, and Peter Burke. **New World Blackbirds: The Ictarids**. Princeton, NJ: Princeton University Press, 1999. 431p. $49.50. ISBN 0691006806.

103 species; color illustrations; maps. Includes bropendolas, blackbirds, orioles, and meadowlarks.

997. Kaufman, Kenn. **A Field Guide to Advanced Birding: Birding Challenges and How to Approach Them**. Boston: Houghton Mifflin, 1990. 299p. (Peterson Field Guide Series, No. 39). $18.00pa. ISBN 0395535174; 0395533767pa.

Covers 35 difficult groups, with emphasis on varying plumage, ages, etc.; black-and-white illustrations. Illustrated by the author.

998. Kaufman, Kenn. **Lives of North American Birds**. Boston: Houghton Mifflin, 1996. 675p. (Peterson Natural History Companions). $35.00. ISBN 0395783224pa; 0395770173.

900 species; color photos; maps; natural history. This is a natural history covering the habitat, feeding, nesting, and migration of North American birds; while there are photos for most of the species covered, there are no descriptions. The book is unique because it is the print version of the CD-ROM *Peterson Multimedia Guides: North American Birds* (see entry 1018) rather than the CD-ROM being a version of the book.

999. Klutz Press. **Everybody's Everywhere Backyard Bird Book**. Palo Alto, CA: Klutz, 1992. 87p. $10.95pa. ISBN 1878257269pa.

26 common species; color photos; natural history. Includes Audubon birdcall. For children and beginners who might not be able to identify a robin. Spiralbound.

1000. Kondo, Riki H. **Instant Nature Guide: Birds**. New York: Grosset and Dunlap, 1979. 48p. ISBN 0448126729.

100 species; black-and-white and color illustrations; maps; brief natural history. Includes four dials with color illustrations of birds and page numbers for quick recognition. Illustrated by Dorothea and Sy Barlowe.

1001. Lambert, Mike, and Alan Pearson. **An Instant Guide to Birds: Nearly 200 of the Most Common North American Birds Described and Illustrated in Full Color**. New York: Crescent, 1985. 128p. (Instant Guide Series). ISBN 0517468913.

200 species; color illustrations; by habitat; includes checklist of the birds covered in the guide.

1002. Lambert, Mike, and Alan Pearson. **An Instant Guide to Freshwater Birds**. New York: Bonanza, 1989. 123p. (Instant Guide Series). $3.99. ISBN 0517667924.

    Not seen; see series entry 48.

1003. Lambert, Mike, and Alan Pearson. **An Instant Guide to Owls and Birds of Prey**. New York: Bonanza, 1989. 124p. (Instant Guide Series). ISBN 051765525X.

    50 species; color illustrations; maps. Descriptions include flight, call, and similar species.

1004. Lambert, Mike, and Alan Pearson. **An Instant Guide to Seabirds**. New York: Bonanza, 1988. 123p. (Instant Guide Series). $3.99. ISBN 0517655241.

    116 species; color illustrations; maps; checklist. Descriptions include similar species.

1005. Le Master, Richard. **Waterfowl Identification: The LeMaster Method**. Chicago: Contemporary, 1986. 75p. $7.95pa. ISBN 0809248484; 0809249421pa.

    43 species; color illustrations of bills, heads, wings, and feet as well as silhouettes. The LeMaster method involves identifying birds in hand by their bills and birds in flight by the usual field marks, plus flight level.

1006. Miller, Millie. **Early Birds: Common Backyard Birds**. Boulder, CO: Johnson, 1997. 64p. (Millie & Cyndi's Pocket Nature Guides). ISBN 1555662056pa.

    "Previously published as Early bird: eastern backyard birds and Early bird: western backyard birds"—T.p.

1007. Miller, Millie, and Cyndi Nelson. **Hummers: Hummingbirds of North America**. Boulder, CO: Johnson, 1987. 32p. (Millie and Cyndi's Pocket Nature Guides). $5.95pa. ISBN 1555660126.

    20 species; color illustrations; maps. Small (4x6 inches); attractively calligraphed.

1008. Miller, Millie, and Cyndi Nelson. **Talons: North American Birds of Prey**. Boulder, CO: Johnson, 1990. 40 leaves. (Millie and Cyndi's Pocket Nature Guides). $5.95pa. ISBN 1555660355pa.

    40 species; color illustrations. Small (4x6 inches); attractively calligraphed.

1009.	Morin, Jim. **Jim Morin's Field Guide to Birds**. New York: Quill, 1985. 64p. ISBN 0688042007pa.

A humorous guide to the birds. The author has made illustrations of 60 genuine species, the way their common names sound (e.g., the tree swallow is swallowing a tree, the red knot is tied up in knots, and the limpkin is on crutches). Most are North American birds. Illustrated by the author.

1010.	Morris, Arthur. **Shorebirds: Beautiful Beachcombers**. Minocqua, WI: NorthWord, 1996. 159p. (Wildlife Series). $14.95pa. ISBN 1559715677pa.

49 species; color photos; natural history; and birding tips.

1011.	National Audubon Society. **National Audubon Society Interactive CD-ROM Guide to North American Birds**. New York: Knopf New Media, 1996. $56.95. ISBN 0679760164. (Windows or Macintosh).

Covers 723 birds, including 700 calls, videos, 2,100 color photographs, 700 maps, texts, and trip planners. Includes text from 13 National Audubon Society field guides. Bird Identifier section allows users to narrow down species by life zone, season/location, shape, color, and size. Also permits users to compile their own life lists.

1012.	National Audubon Society. **Videoguide to the Birds of North America**. Directed, written and narrated by Michael Godfrey. New York: MasterVision, 1985, 1988. 5 videocassettes (VHS and Beta). (Master Vision How-To Series). $29.95 per video or $123.95 for the set.

Volume 1 covers 116 species, from loons to game birds; volume 2 covers 105 species of water birds; volume 3 covers 77 species from doves to woodpeckers; volume 4 covers 98 species of songbirds; and volume 5 covers the remaining 109 species of song birds.

1013.	National Geographic Society. **Field Guide to the Birds of North America**. 3rd ed. Washington, DC: National Geographic Society, 1999. 464p. $21.95pa. ISBN 0792274512pa.

Comprehensive; color illustrations; maps. One of the major North American bird guides.

1014.	National Geographic Society. **Guide to Bird Sounds**. Ithaca, NY: Cornell Laboratory of Onithology, 1985. 2 sound cassettes and booklet. $19.95.

For use with the National Geographic Society's *Field Guide to the Birds of North America* (see entry 1013). Covers 179 species, 120 of them eastern birds.

1015.   Perkins, Simon. **Familiar Birds of Sea and Shore**. New York: Knopf, 1994. 191p. (National Audubon Society Pocket Guides). ISBN 0679749217pa.
        80 species; color photos; maps.

1016.   Peterson, Roger Tory. **How to Know the Birds: An Introduction to Bird Recognition**. 2nd ed. Boston: Houghton Mifflin, 1962. 168p.
        300 species; black-and-white and color illustrations and silhouettes. A beginner's guide with sections on how to identify birds, families of birds (including common species), habitats and the birds found in them, and silhouettes of common species. While intended as a how-to book, it can be used as a field guide. Color illustrations taken from the Wildlife Conservation Stamp Series; black-and-white illustrations by Roger Tory Peterson.

1017.   Peterson, Roger Tory. **Peterson First Guide to Birds of North America**. Boston: Houghton Mifflin, 1986. 128p. $5.95pa. ISBN 039590660pa.
        188 species; color illustrations. Illustrated by Roger Tory Peterson.

1018.   Peterson, Roger Tory. **Peterson Multimedia Guides: North American Birds** (CD-ROM). Boston: Houghton Mifflin Interactive, 1995. $69.95. ISBN 0395730562.
        1,000 species; includes text and illustrations from the Peterson Guides *Eastern Birds* and *Western Birds*, plus life histories, songs, photographs, range maps; interactive video on how to observe birds hosted by Roger Tory Peterson; checklist and glossary. The life history material has been excerpted and published as *Lives of North American Birds*, by Kenn Kaufman (see entry 998).

1019.   Peterson, Roger Tory, and Edward L. Chalif. **Aves de Mexico: Guía de Campo: Identificación de Todas las Especies Encontradas en Mexico, Guatemala, Belice y El Salvador**. 1a ed. Mexico: Editorial Diana, 1989. 473p. ISBN 9681312821.
        Translation of *A Field Guide to Mexican Birds* (see entry 1020).

1020.   Peterson, Roger Tory, and Edward L. Chalif. **A Field Guide to Mexican Birds: Field Marks of All Species Found in Mexico, Guatemala, Belize (British Honduras), El Salvador**. Boston:

Houghton Mifflin, 1973. 298p. (Peterson Field Guide Series, No. 20). $20.00pa. ISBN 0395171296; 0395483549pa.

1,038 species; color illustrations, some black-and-white. Illustrated by Roger Tory Peterson.

1021. Pyle, Porter. **Identification Guide to North American Birds: A Compendium of Information on Identifying, Ageing, and Sexing "Near-Passerines" and Passerines in the Hand**. Bolinas, CA: Slate Creek, 1997- . ISBN 0961894024 (v. 1).

395 species; color illustrations. Covers doves through weavers. An advanced bird-bander's guide; revision of the author's *Identification Guide to North American Passerines*. Illustrated by Steven N. G. Howell.

1022. Reader's Digest Editors. **North American Wildlife: Birds**. New York: Reader's Digest, 1998. 216p. $16.95pa. ISBN 0762100362pa.

650 species; color illustrations; maps.

1023. Reed, Chester Albert. **North American Birds Eggs**. Rev. ed. New York: Dover, 1965. 372p.

768 species; eggs illustrated in life-sized black-and-white photos, birds in line drawings; some photos of nests also included. Illustrated by the author.

1024. Richards, Alan J. **Shorebirds of North America**. New York: Gallery, 1991. 144p. (American Nature Guides). ISBN 0831769629.

70 species; color photos; maps; space for recording sightings.

1025. Rising, Jim. **A Guide to the Identification and Natural History of the Sparrows of the United States and Canada**. San Diego, CA: Academic, 1996. 365p. $39.95; $19.95pa, ISBN 0125889704; 0125889712pa.

All 62 species; color photos, including various plumages; maps; natural history and descriptions include field identification, voice, habitat, range, and breeding. Illustrated by David Beadle.

1026. Robbins, Chandler S., Bertel Bruun, and Herbert S. Zim. **Birds of North America: A Guide to Field Identification**. Expanded, rev. ed. New York: Golden, 1983. 360p. (Golden Field Guide Series). $14.00pa. ISBN 030737002X; 0307336565pa.

750 species; color illustrations facing species accounts; maps; sonograms of bird songs. Index functions as checklist also. Illustrated by Arthur B. Singer.

1027.   Schorre, Barth. **The Wood Warblers: An Introductory Guide**.
        Austin: University of Texas Press, 1998. 140 p. (Corrie Herring
        Hooks Series, No. 35). $35.00; $17.95pa. ISBN 0292777299;
        0292777302pa.

        49 species; color photos; descriptions include distribution,
        status, and natural history.

1028.   Scott, Virgil E., Keith E. Evans, David R. Patton, and Charles P. Stone.
        Cavity-Nesting Birds of North American Forests. Washington,
        DC: Forest Service, U.S. Department of Agriculture, 1977. 112p.
        (Agriculture Handbook, No. 511).

        85 species; color illustrations; maps; pamphlet; descriptions
        include habitats, nests, and food. Designed for land managers.
        Illustrations by Arthur Singer, taken from the Golden Field
        Guide *Birds of North America*.

1029.   Sill, Ben L., Cathryn P. Sill, and John C. Sill. **Beyond Birdwatching:
        More Than There Is to Know About Birding**. Atlanta, GA:
        Peachtree, 1993. 76p. $9.95pa. ISBN 1561450847pa.

        A spoof of birder's magazines containing very unreliable
        information and illustrations of non-existent birds (see entries
        1030 and 1031 for more species). Color and black-and-white
        illustrations; advertisements. Illustrated by John C. Sill.

1030.   Sill, Ben L., Cathryn P. Sill, and John C. Sill. **A Field Guide to
        Little-Known and Seldom-Seen Birds of North America**.
        Atlanta, GA: Peachtree, 1988. 73p. $9.95pa. ISBN 0934601585pa.

        32 imaginary species; color illustrations; maps; plus glos-
        sary, bibliography, and all the features of a "normal" field guide.
        John Sill has illustrated other genuine field guides, so the unreal
        birds look quite real. Illustrated by John C. Sill.

1031.   Sill, Ben L., Cathryn P. Sill, and John C. Sill. **Another Field Guide
        to Little-Known and Seldom-Seen Birds of North America**.
        Atlanta, GA: Peachtree, 1990. 71p. $9.95pa. ISBN 0934601976pa.

        Another 32 imaginary species; color illustrations. See entry
        1030. Illustrated by John C. Sill.

1032.   Sprunt, Alexander, IV, and Herbert S. Zim. **Gamebirds: A Guide to
        North American Species and Their Habits**. New York: Golden,
        1961. 160p. (A Golden Nature Guide). ISBN 0307635139;
        0307244091pa.

        125 species; color illustrations; maps. Illustrated by James G.
        Irving.

1033. Sutton, Clay, and Richard K. Walton. **North American Birds of Prey**. New York: Knopf, 1994. 191p. (National Audubon Society Pocket Guides). $9.00pa. ISBN 0679749233pa.

56 species; color photos, including birds in flight; maps.

1034. Sutton, Patricia, and Clay Sutton. **How to Spot Hawks and Eagles**. Shelburne, VT: Chapters, 1996. 143p. $24.95; $15.95pa. ISBN 1576300013; 1576300005pa.

24 species; color photos; maps; natural history. About one-half of the book consists of hawk-watching tips and background information.

1035. Sutton, Patricia, and Clay Sutton. **How to Spot an Owl**. Shelburne, VT: Chapters, 1994. 143p. $14.95pa. ISBN 1881527352; 1881527360pa.

19 species; color photos; maps; natural history. About one-half of the book consists of owl-watching tips and background information.

1036. Thompson, Stuart L. **160 Birds to Know**. Agincourt, Ontario: Book Society of Canada, 1970. ISBN 0772510911.

160 species; separate color visual keys to male spring plumage of common species and spring plumage of water and game birds and birds of prey; black-and-white and color illustrations; natural history. In two sections: the first identifies 80 common birds of eastern North America and the second discusses another 80 water and game birds and birds of prey. Descriptions include size, voice, nesting, habitat, range, and food. Common names in English and French.

1037. Todd, Frank S. **Handbook of Waterfowl Identification**. Vista, CA: Ibis, 1997. 128p. $19.95pa. ISBN 0934797145pa.

175 species and subspecies worldwide; color photos; maps. Descriptions in main part of book provide physical description only, but the appendix has tables providing weight, description of eggs, incubation, and length of fledgings. Covers swans, geese, and ducks worldwide. Designed as a companion volume to the author's *Natural History of the Waterfowl*.

1038. Walton, Richard K. **Familiar Birds of Lakes and Rivers**. New York: Knopf, 1994. 191p. (National Audubon Society Pocket Guides). $9.00pa. ISBN 0679749225pa.

80 species; color photos; maps.

1039.   Walton, Richard K. **North American Waterfowl**. New York: Knopf, 1994. 192p. (National Audubon Society Pocket Guides). $9.00pa. ISBN 0679749241pa.

74 species; color photos; maps.

1040.   Walton, Richard K., and Robert W. Lawson. **Backyard Bird Song**. New York: Houghton Mifflin, 1991. 1 cassette or CD, 32 page booklet. (Peterson Field Guide Series, No. 43). $21.95 (cassette); $20.00 (CD). ISBN 0395975271 (cassette); 039597528X (CD).

28 species, arranged by type of song.

1041.   Weidensaul, Scott. **National Audubon Society First Field Guide to Birds**. New York: Scholastic, 1998. 160p. $10.95 (flexi); $17.95pa. ISBN 0590054821 (flexi); 0590054465pa.

50 common species covered in detail, another 125 illustrated and described; color photos; extensive introductory material; children's guide. Includes a separate pocket-sized card illustrating the 50 main species.

1042.   Wheeler, Brian K., and William S. Clark. **Photographic Guide to North American Raptors**. San Diego, CA: Academic, 1995. 250p. $32.00. ISBN 0127455302.

43 species of North American diurnal raptors; numerous photos illustrate variant forms and difficult identification problems for both perched and flying birds. Advanced field guide designed to complement the authors' *The Field Guide to Hawks of North America* (see entry 969).

1043.   Wylie, Stephen R., and Stewart S. Furlong. **Key to North American Waterfowl**. Rev. ed. Chestertown, MD: Schroeder Prints, 1978. 32p. (Permaguide Series). ISBN 0931766001.

60 species; color illustrations, flight and takeoff silhouettes; printed on waterproof pages. Illustrated by Jack A. Schroeder.

1044.   Zim, Herbert Spencer, and Ira N. Gabrielson. **Birds: A Guide to Familiar American Birds: 129 Birds in Full Color**. New York: Golden, 1987. 160p. (Golden Guide). $5.95pa. ISBN 0307640531; 0307244903pa.

129 species illustrated, 121 mentioned briefly; color illustrations; maps. Also includes table of arrival and departure dates and brief natural history for 110 common species. Illustrated by James G. Irving.

Eastern North America

1045. Alsop, Fred J. **Birds of the Smokies**. Gatlinburg, TN: Great Smoky Mountains Natural History Association, 1991. 167p. $9.95pa. ISBN 0937207055pa.

    100 species; color photos; checklist. A colorful souvenir guide.

1046. Bessette, Alan E., et al. **Birds of the Adirondacks: A Field Guide**. Utica, NY: North Country, 1993. 240p. ISBN 0932052940pa.

    180 species; color photos separate from text; arranged by broad categories (water birds, songbirds, etc.).

1047. Blachly, Lou, and Randolph Jenks. **Naming the Birds at a Glance: A Guide to the Eastern Land Birds from South Carolina West to the Rocky Mountains and North to the Arctic**. New York: Bonanza, 1989. 331p. ISBN 051767954X.

    240 species; line drawings; arranged by color and pattern rather than taxonomy. Revised edition of *Land Birds of the Northeast*, 1959. Illustrated by Sheridan Oman.

1048. Bovey, Robin B. **Birds of Calgary**. Rev. and expanded. ed. Edmonton, Alta.: Lone Pine, 1990. 128p. $9.95pa. ISBN 0919433820pa.

    55 species; color illustrations. Illustrated by Ewa Pluciennik.

1049. Bull, John L., Edith Bull, and Gerald Gold. **Birds of North America Eastern Region: A Quick Identification Guide to Common Birds**. New York: Macmillan, 1985. 157p. (Macmillan Field Guides). $13.95pa. ISBN 0020796609pa; 0025182307.

    253 species; color illustrations. Illustrated by Pieter D. Prall.

1050. Bull, John L., and John Farrand, Jr. **The National Audubon Society Field Guide to North American Birds, Eastern Region**. Rev. ed. New York: Knopf, 1994. 797p. $19.00 (flexi). ISBN 0679428526 (flexi).

    508 species; color photos; maps; visual key. Arranged by similarity (e.g. "upland ground birds") , rather than by family.

1051. Chapman, Frank Michler. **Bird-Life: A Guide to the Study of Our Common Birds**. New York: D. Appleton, 1919. 195p.

    Color illustrations; natural history. A classic beginner's guide based on Chapman's *Handbook of Birds of Eastern North America*. Illustrated by Ernest Thompson Seton.

1052. Chapman, Frank Michler. **Handbook of Birds of Eastern North America**. 2nd rev. ed. New York: Dover, 1966. 581p.

A classical, pre-Peterson bird guide with identification often based on specimen in hand; keys; black-and-white and color illustrations. Illustrated by Louis Agassiz Fuertes and others.

1053. Cruickshank, Allan D. **Cruickshank's Pocket Guide to the Birds: Eastern and Central North America**. New York: Dodd, Mead, and Co., 1953. 216p.

350 species; visual key to major groups; black-and-white illustrations and color photos. A classic beginner's guide with extensive information on identifying birds visually and by song. Illustrated by Don Eckelberry.

1054. Elliott, Lang, Donald W. Stokes, and Lillian Q. Stokes. **Stokes Field Guide to Bird Songs, Eastern Region**. New York, NY: Time Warner AudioBooks, 1997. 3 sound cassettes or compact disks and booklet. $29.97 (cassettes); $29.98 (CD). ISBN 1570424829 (cassettes); 1570424837 (CD).

372 species. Companion to *Stokes Field Guide to Birds, Eastern Region* (see entry 1092).

1055. Farrand, John. **Eastern Birds**. New York: McGraw-Hill, 1988. 484p. (An Audubon Handbook). $17.95pa. ISBN 0070199760pa.

460 species; color photos; arranged by similarities (habitat, appearance, etc.). Covers area east of the Rocky Mountains.

1056. Fisher, Chris C. **Birds of Detroit**. Edmonton, Alta.: Lone Pine, 1997. 160p. $9.95pa. ISBN 1551051265pa.

125 species; color illustrations; sighting frequency charts.

1057. Fisher, Chris C. **Ontario Birds**. Edmonton, Alta.: Lone Pine, 1996. 159p. $17.95pa. ISBN 1551050692pa.

125 species; color illustrations; natural history.

1058. Fisher, Chris C. **Prairie Birds**. Edmonton, Alta.: Lone Pine, 1995. 125p. $11.95pa. ISBN 155105051Xpa.

90 species; color illustrations; information on setting up bird feeders. Large print. Covers Alberta, Saskatchewan, Manitoba, Montana, North Dakota, and South Dakota. A beginner's guide.

1059. Fisher, Chris C., and David B. Johnson. **Birds of Chicago: Including NE Illinois and NW Indiana**. Edmonton, Alta.: Lone Pine, 1997. 160p. $9.95pa. ISBN 1551051125pa.

126 species; color illustrations; descriptions include similar species and seasonal occurrence; also includes "Quick ID" box with sizes and brief descriptions.

1060. Gillette, John, and David Mohrhardt. **Coat Pocket Bird Book**. Lansing, MI: Thunder Bay, 1995. 160p. $14.95pa. ISBN 1882376145pa.

76 species; color illustrations; separate checklists for birds listed in book and found in Michigan; natural history. Covers the Great Lakes region. Companion to Ham's *Kitchen Table Bird Book* (see entry 1062), despite the name, this guide is too large for any but the largest pocket. Illustrated by Mohrhardt.

1061. Griggs, Jack L. **All the Backyard Birds: East**. New York: HarperPerennial, 1998. 96p. (An American Bird Conservancy Compact Guide). $4.95pa. ISBN 0062736310pa.

84 species; color illustrations; maps; checklist of birds covered in book. Illustrations taken from *All the Birds of North America*.

1062. Ham, John. **Kitchen Table Bird Book**. Lansing, MI: Thunder Bay, 1995. 160p. $15.95pa. ISBN 1882376153pa.

77 species; color illustrations; natural history. Covers birds found at birdfeeders in the Great Lakes area. Companion to Gillette's *Coat Pocket Bird Book* (see entry 1060). Illustrated by David Mohrardt.

1063. Harrison, Hal H. **A Field Guide to Birds' Nests of 285 Species Found Breeding in the United States East of the Mississippi River**. Boston: Houghton Mifflin, 1975. 257p. (Peterson Field Guide Series, No. 21). $26.00; $19.00pa. ISBN 0395936101; 0395936098pa.

285 species; color photos and line drawings. Illustrated by Ned Smith and Mada Harrison.

1064. Hoffmann, Ralph. **A Guide to the Birds of New England and Eastern New York Containing A Key for Each Season and Short Descriptions of over Two Hundred Fifty Species with Particular Reference to Their Appearance in the Field**. Boston: Houghton Mifflin, 1904. 357p.

250 species; seasonal keys; black-and-white illustrations. Another antique guide, interesting because of the keys. One of Roger Tory Peterson's favorite guides as a youngster, in part due to the illustrations by Louis Agassiz Fuertes.

1065. Hogan, Geoff. **Familiar Birds of Prince Edward Island**. Charlottetown, P.E.I.: Ragweed, 1991. 150p. (Island Pathways). ISBN 0921556160.

125 species; color photos; checklist.

1066. Jardine, Ernie. **Bird Song Identification Made Easy**. Toronto: Natural Heritage/Natural History, 1996. 207p. ISBN 189621911Xpa.

125 species; black-and-white illustrations; eastern North America. This is an attempt to classify bird songs so that amateurs can identify them. The songs are arranged into four main categories (very short songs, repeated notes, short songs–varying notes, and long songs–varying notes), and then further subdivided. Each species account includes a sonogram-like illustration of the song, a description of the song, habitat, and seasonal occurrence. Not all species are illustrated. The guide also includes a Habitat Guide (birds arranged by habitat, not type of song), a narrative description of the birds that would be found in those habitats, and 25 other species not covered in the main section. Illustrated by Don Cavin.

1067. Kale, H. W., and David S. Maehr. **Florida's Birds: A Handbook and Reference**. Sarasota, FL: Pineapple, 1990. 288p. $26.95; $19.95pa. ISBN 0910923671; 091092368Xpa.

325 species; color and black-and-white illustrations; index/checklist; background information, including birding sites. Illustrated by Karl Karalus.

1068. Keller, Charles E., and Timothy C. Keller. **Birds of Indianapolis: A Guide to the Region**. Bloomington: Indiana University Press, 1993. 145p. $25.00; $8.95pa. ISBN 0253285348pa, 0253331196.

125 species; color photos separate from text; natural history.

1069. Kidwell, Al. **Coastal Birds: A Guide to Birds of Maine's Beautiful Coastline**. Freeport, ME: DeLorme, 1983. 48p. (A Maine Geographic book.) ISBN 0899330525pa.

49 species; color illustrations; natural history; pamphlet. Illustrated by Jon Luoma.

1070. Lederer, Roger J. **Bird Finder: A Guide to Common Birds of Eastern North America**. Berkeley, CA: Nature Study Guild, 1990. 60p. $3.00pa. ISBN 019255019Xpa.

60 species; black-and-white illustrations; pocket sized. Arranged by habitat. Illustrated by Roger C. Franke.

1071. Mathews, Ferdinand Schuyler. **Field Book of Wild Birds and Their Music: A Description of the Character and Music of Birds, Intended to Assist in the Identification of Species Common in the United States East of the Rocky Mountains**. New York: Dover, 1967. 325p.

    127 species; color illustrations; emphasis on birdsong, includes musical notation of songs. Illustrated by the author.

1072. McKeating, Gerald. **Birds of Ottawa and Vicinity**. Edmonton, Alta.: Lone Pine, 1990. 144p. $9.95pa. ISBN 0919433642pa.

    Not seen; see series entry 11. Illustrated by Ewa Pluciennik.

1073. McKeating, Gerald. **Birds of Toronto and Vicinity**. Edmonton, Alta.: Lone Pine, 1990. 144p. $9.95pa. ISBN 0919433634pa.

    90 species; color illustrations. Illustrated by Ewa Pluciennik.

1074. Melius, M. **Under Wing and Sky**. Rapid City, SD: Fenske Companies, 1995. 88p. $8.95pa. ISBN 0912410132pa.

    215 species; color photos; checklist. Alternate title *Birds of the Badlands and Black Hills*.

1075. Miller, Millie, and Cyndi Nelson. **Early Bird: Eastern Backyard Birds**. Boulder, CO: Johnson, 1990. 40 leaves. (Millie and Cyndi's Pocket Nature Guides). $5.95pa. ISBN 1555660665pa.

    50 species; color illustrations. Small (4x6 inches); attractively calligraphed.

1076. Petersen, Wayne R. **Songbirds and Familiar Backyard Birds (Eastern Region)**. New York: Knopf, 1994. 192p. (National Audubon Society Pocket Guides). $9.00pa. ISBN 0679749268pa.

    80 species; color photos; maps.

1077. Peterson, Roger Tory. **Eastern/Central Bird Songs**. 3rd ed. Boston: Houghton Mifflin, 1990. (Peterson Field Guide Series). Compact disk or sound cassettes. $39.95 (CD); $25.00 (2 cass.); $19.95 (2 LP). ISBN 0395502578 (CD); 0395531500 (2 cass.); 0395346746 (2 LP).

    250 species; arranged to accompany the fourth edition of Peterson's *A Field Guide to the Birds* (see entry 1078).

1078. Peterson, Roger Tory. **A Field Guide to the Birds: A Completely New Guide to All the Birds of Eastern and Central North America**. 4th ed., completely rev. and enl. Boston: Houghton Mifflin, 1980. 384p. (Peterson Field Guide Series, No. 1). $27.00;

$18.00pa; $16.95 (flexi). ISBN 0395911753; 0395911761pa; 039591177X (flexi).

389 species, plus 119 accidentals; color illustrations; maps; checklist; includes description, similar species, voice, and range. Text and illustrations are integrated and maps are in a separate section. Illustrated by Roger Tory Peterson.

1079.   Peterson, Roger Tory. **A Field Guide to the Birds: Commemorative Edition**. Boston: Houghton Mifflin, 1996. 264p. $18.95. ISBN 0395854938.

A commemorative reprint of Peterson's original 1934 field guide. Has primarily black-and-white illustrations, except for the small, colorful songbirds. Covers about 450 species plus many subspecies (more than are currently recognized). Some very common species, such as the common crow, robin, and English sparrow, are merely mentioned with no description or illustration because "everyone knows what they look like." Many features of Peterson's later guides are found in this first edition, but due to the many changes in nomenclature since 1934, it should not be used as a working field guide. Illustrated by Roger Tory Peterson.

1080.   Peterson, Roger Tory. **A Field Guide to the Birds of Texas and Adjacent States**. Boston: Houghton Mifflin, 1963. 304p. (Peterson Field Guide Series, No. 13). $21.00pa. ISBN 0395921384pa.

487 species plus 55 accidentals; color and black-and-white illustrations; 66 bird silhouettes; checklist. Illustrated by Roger Tory Peterson.

1081.   Potter, Eloise F., James F. Parnell, Robert P. Teulings. **Birds of the Carolinas**. Chapel Hill: University of North Carolina Press, 1980. 408p. $19.95pa. ISBN 0807841552pa.

415 species; color photos; background information; descriptions include range, nesting, and feeding.

1082.   Pough, Richard Hooper. **Audubon Bird Guide: Eastern Land Birds**. Garden City, NY: Doubleday, 1946. 312p.

275 species; color illustrations; background information and natural history. Covers area east of the 100th meridian. Illustrated by Don Eckelberry.

1083.   Pough, Richard Hooper. **Audubon Guides: All the Birds of East and Central North America**. Garden City, NY: Doubleday, 1953. 312, 352p.

533 species; color illustrations; separate sections on small land, water, game, and large land birds. More an extensive natural history than most field guides, this book includes habits, voice, range, and nesting. Illustrated by Don Eckelberry and Earl L. Poole.

1084.   Pough, Richard Hooper. **Audubon Land Bird Guide: Small Land Birds of Eastern and Central North America from Southern Texas to Central Greenland**. Garden City, NY: Doubleday, 1963. 312p. (Doubleday Nature Guides Series).

960 species; color illustrations separate from text; natural history including habits, voice, range, and nesting. Illustrated by Don Eckelberry.

1085.   Pough, Richard Hooper. **Audubon Water Bird Guide: Water, Game and Large Land Birds, Eastern and Central North America, from Southern Texas to Central Greenland**. Garden City, NY: Doubleday, 1956. 352p.

258 species; color illustrations separate from text, black-and-white illustrations integrated with text; natural history. More an extensive natural history than most field guides, this book includes habits, voice, range, and nesting. The subtitle of this field guide is more accurate than the title, since the guide actually covers the bird families from loons to pigeons (the most "primitive" birds). Illustrated by Don Eckelberry and Earl L. Poole.

1086.   Rappole, John H., and Gene W. Blacklock. **Birds of Texas: A Field Guide**. College Station: Texas A&M University Press, 1994. 280p. (W. L. Moody, Jr., Natural History Series, No. 14). $16.95pa; $34.95. ISBN 0890965455pa; 0890965447.

622 species; color photos, one per species; more information on habitat and range than the usual field guide.

1087.   Reed, Chester Albert. **Bird Guides**. New York: Doubleday, Page and Co., 1909. 2 v. Contents: Pt. 1. **Water Birds, Game Birds and Birds of Prey, East of the Rockies**. Pt. 2. **Land Birds East of the Rockies from Parrots to Bluebirds**.

Part 2 contains 766 species; color illustrations; natural history, including information on song, nest, and range; pocket sized. A classic; this is the guide Roger Tory Peterson used before developing his own field guide. Illustrated by the author.

1088.   Robison, B. C. **Birds of Houston**. Houston, TX: Rice University Press, 1990. 129p. $22.50; $12.95pa. ISBN 0892633034; 0892633042pa.

55 species; color photos; natural history; beginner's guide.

1089. Rorimer, Irene Turk. **A Field Key to Our Common Birds**. Cleveland, OH: Cleveland Museum of Natural History, 1940. 160p. (Cleveland Museum of Natural History Pocket Natural History, Zoological Series, No. 3).

      250 species; key arranged by habitat, size, color; color and black-and-white illustrations. Covers northeastern United States with special emphasis on northern Ohio. Illustrated by Roger Tory Peterson.

1090. Shaw, Frank. **Eastern Birds**. New York: Gallery, 1990. 232p. (American Nature Guides). ISBN 0831769653.

      430 species; color illustrations; spiralbound.

1091. Stokes, Donald W., and Lillian Q. Stokes. **Stokes Beginner's Guide to Birds: Eastern Region**. Boston: Little, Brown and Co., 1996. 120p. $7.95pa. ISBN 0316818119.

      100 species; color photos; maps; arranged by color and size. Includes description, voice, habitat, nest, and how to attract birds.

1092. Stokes, Donald W., and Lillian Stokes. **Stokes Field Guide to Birds, Eastern Region**. Boston: Little, Brown, 1995. 471p. $16.95pa. ISBN 0316818097pa.

      450 species; color photos; maps; description includes habitat, range, behavior, conservation trends. Includes quick guide to 50 of the most common species and "learning pages" for difficult groups; each species is covered on one page. Also published in Canada as *Guide d'Identification des Oiseaux de l'Est de l'Amerique du Nord*.

1093. Sutton, George Miksch. **Fifty Common Birds of Oklahoma and the Southern Great Plains**. Norman: University of Oklahoma Press, 1977. 113p. $22.95; $15.95pa. ISBN 0806114398; 0806117044pa.

      Color illustrations; natural history. Illustrated by the author.

1094. Trull, Peter. **A Guide to the Common Birds of Cape Cod**. Brewster, MA: Cape Cod Museum of Natural History, 1991. 72p. $8.95pa. ISBN 0916275051pa.

      120 species; black-and-white illustrations; includes three birding tours of Cape Cod. A beginner's guide. Illustrated by Kathy Clark.

1095. Waldon, Bob. **Feeding Winter Birds**. Stillwater, MN: Voyageur, 1991. 190p. $14.95pa. ISBN 0896581667pa.

   47 species of birds likely to be found at birdfeeders in the Midwest; black-and-white illustrations; natural history. Unlike most birdfeeder guides, the emphasis here is less on how or what to feed than on which birds will be found at a feeder. Illustrated by Peter Sawatzky.

1096. Walton, Richard K., and Robert W. Lawson. **Birding by Ear**. Boston: Houghton Mifflin, 1989. 3 cassettes or CDs plus 64p. booklet. (Peterson Field Guide Series, No. 38). ISBN 0395500877 (cassettes), $35.00; 0395712580 (CDs), $35.00.

   Covers 92 species; arranged by type of call. Title on container: *Eastern/Central Birding by Ear: A Guide to Bird Song Identification*.

1097. Walton, Richard K., and Robert W. Lawson. **More Birding by Ear: A Guide to Bird Song Identification, Eastern and Central North America**. Boston: Houghton Mifflin, 1994. 3 cassettes or CDs plus 64p. booklet. (Peterson Field Guide Series, No. 47). ISBN 0395712602 (3 cassettes), $35.00; 0395712599 (3 CDs), $35.00.

   Covers 90 species not included in either Eastern/Central or Western *Birding by Ear*. Arranged by type of call.

1098. Wauer, Roland H. **A Field Guide to Birds of the Big Bend**. 2nd ed. Houston, TX: Gulf, 1996. 290p. (Gulf's Field Guide Series). $18.95pa. ISBN 0877192715pa.

   450 species; black-and-white illustrations; natural history; intended as a supplement to standard field guides of the area. Covers where and when to find birds in Big Bend. Illustrated by Nancy McGowan.

1099. Whitman, Ann, ed. **Familiar Birds of North America: Eastern Region**. New York: Knopf, 1986. 192p. (Audubon Society Pocket Guides). $9.00pa. ISBN 0394748395pa.

   80 species; color photos; maps; covers area west to the Rocky Mountains.

Western North America

1100. Armstrong, Robert H. **Alaska's Birds: A Guide to Selected Species**. Anchorage: Alaska Northwest, 1994. 124p. (Alaska Pocket Guide). $14.95pa. ISBN 0882404555pa.

   80 representative species; color photos, checklist, birding sites.

1101. Armstrong, Robert H. **Guide to the Birds of Alaska**. 4th ed. Anchorage, AK: Alaska Northwest, 1995. 330p. $24.95pa. ISBN 0882404628pa.

All 443 species; color photos; includes field marks, similar species, habitat; checklist.

1102. Baron, Nancy, and John Harrison Acorn. **Birds of the Pacific Northwest Coast**. Renton, WA: Lone Pine, 1997. 240p. $15.95pa. ISBN 155105082Xpa.

250 species; color illustrations. Also published as *Birds of Coastal British Columbia*.

1103. Beedy, Ted, and Steve Granholm. **Discovering Sierra Birds**. Yosemite Association, 1985. 175p. (Discovering Sierra Series). $9.95pa. ISBN 0939666421pa.

180 species; color illustrations; natural history.

1104. Berger, Andrew John. **Bird Life in Hawaii**. 6th ed. Honolulu, HI: Island Heritage, 1987. 72p. ISBN 0896100901pa.

64 species; color illustrations and photos; natural history. Common names in English and Hawaiian.

1105. Bovey, Robin B. **Birds of Edmonton**. Rev. ed. Edmonton, Alta.: Lone Pine, 1990. 128p. $9.95pa. ISBN 0919433804pa.

75 species; color illustrations. Illustrated by Ewa Pluciennik.

1106. Bovey, Robin B., and R. Wayne Campbell. **Birds of Vancouver and the Lower Mainland**. Edmonton, Alta.: Lone Pine, 1989. 143p. $9.95pa. ISBN 0919433731pa.

90 species; color illustrations. Illustrated by Ewa Pluciennik.

1107. Bovey, Robin B., R. Wayne Campbell, and Bryan R. Gates. **Birds of Victoria and Vicinity**. Edmonton, Alta.: Lone Pine, 1989. 143p. $9.95pa. ISBN 0919433758pa.

75 species; color illustrations. Illustrated by Ewa Pluciennik.

1108. Brown, Vinson, and Henry G. Weston, Jr. **Handbook of California Birds**. 3rd ed. Happy Camp, CA: Naturegraph, 1986. 224p. $12.95pa. ISBN 0911010173; 0911010165pa.

368 species; color and black-and-white illustrations; checklist; natural history and background information. Illustrated by Jerry Buzzell.

1109. Bull, John L., and Edith Bull. **Birds of North America: Western Region: A Quick Identification Guide to Common Birds**. New York: Macmillan, 1989. 144p. (Macmillan Field Guides). $12.95pa. ISBN 0020625804pa.

    340 of the most common Western species. Illustrated by James Coe.

1110. Clarke, Herbert. **An Introduction to Northern California Birds**. Missoula, MT: Mountain, 1995. 189p. $14.00pa. ISBN 0878423125pa.

    202 species; color photos; arranged by habitat; natural history. Beginner's guide. Covers area south to Santa Barbara County.

1111. Clarke, Herbert. **An Introduction to Southern California Birds**. Missoula, MT: Mountain, 1989. 186p. $14.00pa. ISBN 0878422331pa.

    215 species; color photos; arranged by habitat; natural history. Beginner's guide.

1112. Cunningham, Richard L. **50 Common Birds of the Southwest**. Tucson, AZ: Southwest Parks and Monuments Association, 1990. 60p. $5.95pa. ISBN 0911408835pa.

    Color photos; natural history; common names in English and Spanish; covers Arizona, New Mexico, and western Texas.

1113. Davis, Barbara L. **A Field Guide to the Birds of the Desert Southwest**. Houston, TX: Gulf, 1997. 360p. (Gulf's Field Guide Series). $21.95pa. ISBN 0884152782pa.

    200 species described, 123 illustrated; color photos; 21 birding sites; covers birds of the Mojave, Sonoran, and Chihuahuan deserts. Works best as a supplementary guide because the photos are in separate plates in the center of the book and do not refer back to the species accounts, which are full of detail on description, voice, habitat, and behavior.

1114. Cogswell, Howard L. **Water Birds of California**. Berkeley, CA: University of California Press, 1977. 399p. (California Natural History Guides, No. 40). $12.95pa. ISBN 0520029991; 0520026993pa.

    200 species; black-and-white and color illustrations; extensive natural history. Illustrated by Gene Christman.

1115. Farrand, John. **Western Birds**. New York: McGraw-Hill, 1988. 477p. (An Audubon Handbook). $17.95pa. ISBN 0070199779pa.

    460 species; color photos; arranged by similarities (habitat, appearance, etc.); advanced guide. Covers area west of Rocky Mountains.

1116. Fisher, Chris C. **Birds of Seattle and Puget Sound**. Vancouver, B.C.: Lone Pine, 1996. 159p. $9.95pa. ISBN 1551050781pa.

125 species; color illustrations; checklist.

1117. Fisher, Chris C. **Birds of the Rocky Mountains**. Edmonton, Alta.: Lone Pine, 1997. 336p. $19.95pa. ISBN 1551050919pa.

320 species; color illustrations; descriptions include voice, distribution, natural history. Includes a 10-page comparison chart showing all birds covered.

1118. Fisher, Chris C. **West Coast Birds**. Edmonton, Alta.: Lone Pine, 1996. 125p. $11.95pa. ISBN 1551050498pa.

90 species; color illustrations. Large print. A beginner's guide.

1119. Fisher, Chris C., and Herbert Clarke. **Birds of San Diego**. Edmonton, Alta.: Lone Pine, 1997. 159p. $9.95pa. ISBN 1551051028pa.

Not seen; see series entry 11.

1120. Fisher, Chris C., and Joseph Morlan. **Birds of San Francisco and the Bay Area**. Edmonton, Alta.: Lone Pine, 1996. 159p. $9.95pa. ISBN 1551050803pa.

Not seen; see series entry 11.

1121. Fisher, Chris C., Ted Nordhagen, and Gary Ross. **Birds of Los Angeles**. Edmonton, Alta.: Lone Pine, 1997. 160p. $9.95pa. ISBN 1551051044pa.

Not seen; see series entry 11.

1122. Fisher, Chris C., Gary Ross, and Greg Butcher. **Birds of Denver and the Front Range**. Edmonton, Alta.: Lone Pine, 1997. 159p. $9.95pa. ISBN 1551051060pa.

Not seen; see series entry 11.

1123. Follett, Dick. **Birds of Crater Lake National Park**. Crater Lake: Crater Lake Natural History Association, 1979. 80p.

75 species; color photos; checklist; natural history.

1124. Follett, Dick. **Birds of Yellowstone and Grand Teton National Parks**. Boulder, CO: Roberts Rinehart, 1975. 73p. ISBN 0911797149.

64 species; color photos and black-and-white illustrations; checklist; natural history. Includes habitat guide for the area.

1125. Gray, Mary Taylor. **Watchable Birds of the Rocky Mountains**. Missoula, MT: Mountain, 1992. 155p. $14.00pa. ISBN 0878422811pa.

　　　70 species; color photos; arranged by habitat; natural history; for the casual or beginning bird-watcher.

1126. Gray, Mary Taylor. **Watchable Birds of the Southwest**. Missoula, MT: Mountain, 1995. 187p. $14.00pa. ISBN 0878423222pa.

　　　68 species; color photos; natural history; arranged by habitat. For the beginning bird-watcher.

1127. Griggs, Jack L. **All the Backyard Birds: West**. 1st ed. New York: HarperCollins, 1998. 96p. (An American Bird Conservancy Compact Guide). ISBN 0062736329.

　　　90 species; color illustrations; maps; checklist of birds covered in book. Illustrations taken from *All the Birds of North America*.

1128. Harrison, Hal H. **A Field Guide to Western Birds' Nests of 520 Species Found Breeding in the United States West of the Mississippi River**. Boston: Houghton Mifflin, 1979. 279p. (Peterson Field Guide Series, No. 25). $17.95pa. ISBN 0395276292; 0395478634pa.

　　　520 species; color and black-and-white photos.

1129. Hawaii Audubon Society. **Hawaii's Birds**. 4th ed. Honolulu, HI: The Society, 1989. 112p. $9.95pa. ISBN 1889708003pa.

　　　141 species; color photos, some color illustrations; includes natural history; pamphlet. First edition published in 1959 as *Hawaiian Birds*.

1130. Holroyd, Geoffrey L. **The Compact Guide to Birds of the Rockies**. Edmonton, Alta.: Lone Pine, 1990. 143p. $7.95pa. ISBN 0919433529pa.

　　　110 species; color illustrations; checklist; includes "Brute Force Index" consisting of all species illustrated on four adjacent pages. Each species is illustrated in its natural habitat and has interesting tidbits of natural history. Covers Canadian species. Illustrated by Howard Coneybeare.

1131. Larrison, Earl Junior. **Birds of the Pacific Northwest: Washington, Oregon, Idaho, and British Columbia**. Moscow, ID: University Press of Idaho, 1981. 337p. (Northwest Naturalist Books). ISBN 0893010782pa.

Comprehensive; black-and-white illustrations and photos; more of a natural history guide than identification guide. Revised edition of *Washington Birds*.

1132. Lederer, Roger J. **Pacific Coast Bird Finder: A Manual for Identifying 61 Common Birds of the Pacific Coast**. Berkeley, CA: Nature Study Guild, 1977. $3.00pa. ISBN 091255004Xpa.

61 species; key throughout; black-and-white illustrations; arranged by location/habitat; pocket sized.

1133. McEneaney, Terry. **Birds of Yellowstone: A Practical Habitat Guide to the Birds of Yellowstone National Park—and Where to Find Them**. Boulder, CO: Roberts Rinehart, 1988. 171p. $10.95pa. ISBN 0911797440pa.

20 representative species described and illustrated; all species listed by habitat; color photos; maps; checklist.

1134. Miller, Millie, and Cyndi Nelson. **Early Bird: Western Backyard Birds**. Boulder, CO: Johnson, 1991. 40 leaves. (Millie and Cyndi's Pocket Nature Guides). $5.95pa. ISBN 1555660754pa.

54 species; color illustrations. Small (4x6 inches); attractively calligraphed.

1135. Nehls, Harry B., and R. Bruce Horsfall. **Familiar Birds of the Northwest Covering Birds Commonly Found in Oregon, Washington, Idaho, Northern California, and Western Canada**. 3rd ed. Portland, OR: Portland Audubon Society, 1990. 184p. $11.95. ISBN 0931686083.

350 species; color illustrations; natural history. Updated edition of the author's *Familiar Birds of Northwestern Shores and Waters* and Marshall's *Familiar Birds of Northwestern Forests, Fields, and Gardens*.

1136. Paulson, Dennis R. **Shorebirds of the Pacific Northwest**. Seattle, WA: University of Washington Press: Seattle Audubon Society, 1993. 406p. $27.50pa. ISBN 0295972335; 029597706Xpa.

61 species; color photos; a field guide supplement covering natural history, and advanced identification problems; small enough to carry into the field.

1137. Peterson, Roger Tory. **A Field Guide to Western Bird Songs: Western North American and the Hawaiian Islands**. Boston, Houghton Mifflin, 1975. (Peterson Field Guide Series). Compact disk or sound cassettes. $39.95 (CD); $35.00 (cassettes); $19.95

(3 LPs). ISBN 039571451 (CD); 039551746X (3 cass.); 0395080894 (3 LPs).

522 species; arranged to accompany Roger Tory Peterson's *A Field Guide to Western Birds*, 2nd ed.

1138. Peterson, Roger Tory. **A Field Guide to Western Birds: A Completely New Guide to Field Marks of All Species Found in North America West of the 100th Meridian and North of Mexico.** 3rd ed., completely rev. and enl. Boston: Houghton Mifflin, 1990. 432p. (Peterson Field Guide Series, No. 2). $26.00; $18.00pa. ISBN 0395911745; 0395911737pa.

700 species; color illustrations; maps. Illustrated by Roger Tory Peterson.

1139. Pough, Richard Hooper. **Audubon Western Bird Guide; Land, Water, and Game Birds, Western North America, Including Alaska, from Mexico to Bering Strait and the Arctic Ocean.** Garden City, NY: Doubleday, 1957. 316p.

614 species, 203 covered in detail; black-and-white and color illustrations; natural history including habits, voice, range, and nesting. Illustrated by Don Eckelberry and Terry M. Shortt.

1140. Pratt, H. Douglas, Phillip L. Bruner, and Delwyn G. Berrett. **A Field Guide to the Birds of Hawaii and the Tropical Pacific.** Princeton, NJ: Princeton University Press, 1987. 409p. $85.00; $39.50pa. ISBN 0691084025; 0691023999pa.

Nearly 1,000 species; color illustrations; maps; checklists. Covers the Hawaiian Islands, Fiji, Samoa, Tonga, Southeast Polynesia, and Micronesia. Illustrated by H. Douglas Pratt.

1141. Raynes, Bert. **Birds of Grand Teton National Park and the Surrounding Area.** Moose, WY: Grand Teton Natural History Association, 1984. 90p. $9.95pa. ISBN 0931895006pa.

60 species; color photos; maps. Descriptions include where to find birds, comments, and similar species.

1142. Scott, Virgil E., and David R. Patton. **Cavity-Nesting Birds of Arizona and New Mexico Forests.** Fort Collins, CO: Rocky Mountain Forest and Range Experiment Station, Forest Service, U.S. Dept. of Agriculture, 1975. 52p. (General Technical Report, No. 10).

41 species; black-and-white illustrations; includes range, nest, and food. For of Forest Service personnel in deciding timber harvest and creating environmental impact statements.

1143. Scotter, George Wilby, T. J. Ulrich, and E. G. Jones. **Birds of the Canadian Rockies**. Saskatoon: Western Producer Prairie, 1990. 170p. $22.95. ISBN 0888333056.

200 species; color photos; checklist; natural history.

1144. Seacrest, Betty R., and Delbert A. McNew. **Rocky Mountain Birds: Easy Identification**. 2nd ed. Boulder, CO: Avery, 1993. 101p. $7.95pa. ISBN 093732101Xpa.

99 species; color photos; arranged by primary color of male birds; natural history; pocket-sized beginner's guide.

1145. Shaw, Frank. **Western Birds**. New York: Gallery, 1990. 232p. (American Nature Guides). ISBN 0831769661.

430 species; color illustrations; spiralbound.

1146. Stebbins, Cyril A. **Birds of Yosemite**. Rev. ed. Yosemite National Park: Yosemite Association, 1974. 80p. ISBN 0939666006.

200 species; key; black-and-white illustrations, including flight forms of birds of prey; checklist. Natural history, including information on life zones. Illustrated by Robert C. Stebbins.

1147. Stokes, Donald W., and Lillian Q. Stokes. **Stokes Beginner's Guide to Birds: Western Region**. Boston: Little, Brown and Co., 1996. 120p. $7.95pa. ISBN 0316818127pa.

100 species; color photos; maps; arranged by color and size. Includes description, voice, habitat, nest, and how to attract birds.

1148. Stokes, Donald W., and Lillian Stokes. **Stokes Field Guide to Birds, Western Region**. Boston: Little, Brown, 1995. 519p. $16.95pa. ISBN 0316818100pa.

500 species; color photos; maps; description includes habitat, range, behavior, conservation trends. Includes quick guide to 50 of the most common species and "learning pages" for difficult groups; each species is covered on one page.

1149. Stromsem, Nancy E. **A Guide to Alaskan Seabirds**. 2nd ed. Anchorage, AK: Alaska Natural History Association, 1989. 44p. $4.00pa. ISBN 0960287647pa.

40 species; black-and-white illustrations; maps; natural history. Illustrated by Charlotte I. Adamson.

1150. Udvardy, Miklos D. F. **National Audubon Society Field Guide to North American Birds, Western Region**. Rev. ed. New York: Knopf, 1994. 822p. $19.00 (flexi). ISBN 0679428518 (flexi).

544 species; color photos; maps; visual key. Arranged by similarity (e.g. "upland ground birds"), rather than by family. Perching birds are also arranged by color.

1151. Ulrich, Tom J. **Birds of the Northern Rockies**. Missoula, MT: Mountain, 1984. 159p. $12.00pa. ISBN 0878421696pa.

170 species; color photos; very attractive. Covers Rockies from Alberta to Wyoming.

1152. Walton, Richard K. **Songbirds and Familiar Backyard Birds (Western Region)**. New York: Knopf, 1994. 192p. (National Audubon Society Pocket Guides). $9.00pa. ISBN 0443046670; 067974925Xpa.

80 species; color photos; maps.

1153. Walton, Richard K., and Robert W. Lawson. **Birding by Ear: Western**. Boston, MA: Houghton Mifflin, 1990. 3 sound cassettes or compact disks and booklet. (Peterson Field Guide Series, No. 41). $35.00 (cassettes); $35.00 (CDs). ISBN 1395528119 (cassettes); 0395712572 (CDs).

Covers 85 species; arranged by type of call. Title on container: *Western Birding by Ear: A Guide to Bird Song Identification*.

1154. Wassink, Jan L. **Birds of the Central Rockies**. Missoula, MT: Mountain, 1991. 179p. $14.00pa. ISBN 0878422358pa.

191 species; color photos; natural history. Covers Colorado, Wyoming, Utah, and Idaho.

1155. Wassink, Jan L. **Birds of the Pacific Northwest Mountains: The Cascade Range, the Olympic Mountains, Vancouver Island, and the Coast Mountains**. Missoula, MT: Mountain, 1995. 199p. $18.00pa. ISBN 0878423087pa.

197 species; color photos; descriptions include field identification, status, and notes. Very attractive.

1156. Whitman, Ann, ed. **Familiar Birds of North America: Western Region**. New York: Knopf, 1986. 192p. (Audubon Society Pocket Guides). $9.00pa. ISBN 0394748425pa.

80 species; color photos; maps; covers area west of Rocky Mountains.

Chapter 15

Mammals

This chapter contains field guides about mammals, including pets, domestic animals, and marine mammals. Since there are relatively few conspicuous mammal species (compared to birds, for instance), most mammal field guides cover fewer species and have room to cover their natural history. It is not uncommon to find field guides to the tracks and traces of mammals, since so many of them are rarely seen. These guides may include tracks, skulls and bones, scats (droppings), and other objects demonstrating that a mammal has been in the area.

See also Chapter 2, "Flora and Fauna," and Chapter 9, "Animals," for more mammal field guides.

NORTH AMERICA

1157. Alden, Peter. **Peterson First Guide to Mammals of North America**. Boston: Houghton Mifflin, 1987. 128p. $4.95pa. ISBN 0395427673pa.

    200 species; color and black-and-white illustrations; based on Burt's *Field Guide to the Mammals*. Illustrated by Richard P. Grossenheider.

1158. Amos, Stephen H. **Familiar Marine Mammals**. New York: Knopf, 1990. 191p. (Audubon Society Pocket Guides). $9.00pa. ISBN 0679729836pa.

    58 species of North America; color photos; more background information than most Audubon pocket guides.

1159. Boitani, Luigi, and Stefania Bartoli. **Simon and Schuster's Guide to Mammals**. New York: Simon and Schuster, 1983. 511p. (Nature Guide Series). $15.00pa. ISBN 0671454471; 0671428055pa.

     426 species worldwide; color photos and line drawings; maps. Also published in Spain by Grijalbo.

1160. Booth, Ernest S. **How to Know the Mammals**. 4th ed. Dubuque, IA: W. C. Brown Co., 1982. 198p. (Pictured Key Nature Series). $25.80pa. ISBN 0697047814pa.

     400 species; keys; black-and-white illustrations; maps.

     ❑ Boschung, Herbert T., David K. Caldwell, Melba C. Caldwell, Daniel W. Gotshall, and James D. Williams. **The Audubon Society Field Guide to North American Fishes, Whales, and Dolphins**. See entry 710.

1161. Brown, Larry N. **Sea Mammals: Atlantic, Gulf, and Caribbean**. Miami, FL: Windward, 1991. 64p. ISBN 0893170399.

     Black-and-white illustrations, some color photos; about one-half of the book is about natural history. Illustrated by Sandra Romashko.

1162. Bulloch, David K. **The Whalewatcher's Handbook: A Guide to the Whales, Dolphins, and Porpoises of North America**. New York: Lyons & Burford, 1993. $13.95pa. ISBN 1558212329pa.

     Comprehensive; black-and-white illustrations and photos; natural history; includes information on locations, trips, sighting hints, etc. Illustrated by Lou Burlingame.

1163. Burt, William Henry. **A Field Guide to the Mammals: Field Marks of All North American Species Found North of Mexico**. 3rd ed. Boston: Houghton Mifflin, 1976. 289p. (The Peterson Field Guide Series, No. 5). $24.95; $18.00pa. ISBN 0395240824; 0395910986pa.

     380 species; color and black-and-white illustrations, most in separate plate section; maps; tracks, skulls and dental formulae for some species; checklist. Illustrated by Richard P. Grossenheider.

1164. Burton, John A. **Mammals**. New York: Smithmark, 1991. 192p. (American Nature Guides). $9.95. ISBN 0831769734.

     180 species; color illustrations; maps; includes tracks; descriptions include size, identification, range, habitat, food, breeding, conservation, and related species. Also published in the U.K. by Dragon's World. Illustrated by Jim Channell.

1165. Carwardine, Mark. **Whales, Dolphins, and Porpoises**. 1st American ed. New York: Dorling Kindersley, 1995. 256p. (Eyewitness Handbooks). $17.95 (flexi). ISBN 1564586219; 1564586200 (flexi).

100 species; visual key; color illustrations. Illustrated by Martin Camm.

☐ Claridge, Edward, and Betty Ann Milligan. **Animal Signatures**. See entry 735.

☐ DeLorme Publishing. **Wildlife Signatures: A Guide to the Identification of Tracks and Scat**. See entry 737.

1166. Emmons, Louise, and Francois Feer. **Neotropical Rainforest Mammals: A Field Guide**. 2nd ed. Chicago: University of Chicago Press, 1997. 307p. $29.95pa. ISBN 0226207196; 0226207218pa.

315 species, 220 illustrated; color illustrations; maps; natural history. Descriptions include similar species, natural history, range, status, and common names.

1167. Emmons, Louise H., Bret M. Whitney, and David L. Ross, Jr. **Sounds of Neotropical Rainforest Mammals: An Audio Field Guide**. Chicago: Distributed for the Cornell Laboratory of Ornithology by the University of Chicago Press, 1997. Two compact discs (105 minutes) and accompanying booklet. $24.95. ISBN 0938027409.

109 species, including most primates and common large mammals. Booklet includes cites to relevant pages in *Neotropical Rainforest Mammals: A Field Guide* (see entry 1166).

1168. ETI Editors. **Marine Mammals of the World**. Amsterdam, Netherlands: Expert Center for Taxonomic Identification, University of Amsterdam; Springer Electronic Media, 1996. 1 CD-ROM disc plus booklet. (World Biodiversity Database). ISBN 3540145087 (Windows); 3540145095 (Macintosh).

119 species; interactive identification key; color photos; maps; sound recordings. Based on the FAO Species Identification Guide *Marine Mammals of the World*, by Thomas A. Jefferson et al. (see entry 1180).

1169. Farrand, John. **Familiar Animal Tracks**. New York: Knopf, 1993. 191p. (Audubon Society Pocket Guides). $9.00pa. ISBN 0679741488pa.

80 species; color photos; includes tracks and line drawings of scats and other signs.

1170. Farrand, John, ed. **Familiar Mammals**. New York: Knopf, 1987. 192p. (Audubon Society Pocket Guides). $9.00pa. ISBN 0394757963pa.

    80 species illustrated, 30 mentioned; color photos; maps; tracks for most species.

1171. Fichter, George S. **Whales and Other Marine Mammals**. New York: Golden, 1990. 160p. (Golden Guide). $6.95pa. ISBN 0307640754 (Scholar's ed.); 0307240754pa.

    33 whales, 19 seals, plus sea otter, manatee, and polar bear; color illustrations; natural history. Illustrated by Barbara J. Hoopes Ambler.

1172. Forey, Pamela, and Cecilia Fitzsimons. **An Instant Guide to Mammals: The Most Familiar Species of North American Mammals Described and Illustrated in Color**. New York: Bonanza, 1986. 124p. ISBN 0517616769.

    100 species; visual key; color illustrations. Also published by Crescent Books.

1173. Forrest, Louise Richardson. **Field Guide to Tracking Animals in Snow**. Harrisburg, PA: Stackpole, 1988. 192p. ISBN 0811722406.

    55 species; key; black-and-white illustrations; maps. Descriptions include trail pattern, signs (i.e., scats), habitat. Illustrated by Denise Casey.

1174. Graham, Gary L. **Bats of the World: 103 Species in Full Color**. New York: Golden, 1994. 160p. (Golden Guide). $5.50pa. ISBN 0307240800pa.

    Covers 100 of the 1,000 species worldwide; color illustrations. Illustrated by Fiona A. Reid.

1175. Grassy, John, and Chuck I. Keene. **National Audubon Society First Field Guide to Mammals**. New York: Scholastic, 1998. 160p. $17.95; $10.95pa. ISBN 0590054716; 0590054899pa.

    50 common species covered in detail, another 79 illustrated and described; color photos; extensive introductory material; children's guide. Includes a separate pocket-sized card illustrating the 50 main species.

1176. Halfpenny, James C. **A Field Guide to Mammal Tracking in North America**. 2nd ed. Boulder, CO: Johnson, 1986. 161p. $14.95pa. ISBN 0933472986pa.

    Black-and-white illustrations; extensive background and natural history. Illustrated by Elizabeth Blesiot.

1177. Heintzelman, Donald S. **A World Guide to Whales, Dolphins, and Porpoises**. Tulsa, OK: Winchester, 1981. 156p. ISBN 0876913230pa.

    82 species; black-and-white illustrations; checklist; information on whalewatching, sites. Illustrated by Rod Arbogast.

    ❑ Hoffmeister, Donald Frederick. **Zoo Animals**. See entry 718.

1178. Hoyt, Erich. **The Whale Watcher's Handbook**. Garden City, NY: Doubleday, 1984. 208p. ISBN 0385190360pa.

    76 species; black-and-white illustrations; one-half of the book tells where to find whales. Illustrated by Pieter Folkens.

1179. Ingall, Marjorie. **The Field Guide to North American Males**. New York: H. Holt, 1996. 216p. $9.95pa. ISBN 0805042199pa.

    57 types of males; black-and-white illustrations; "natural" history, including diet, nest type, foraging behavior, plumage, etc. Arranged by "family" such as Artsy or Gainfully Employed. "Parodies and pays homage to the National Audubon Society Field Guide to North American Birds" according to the introduction, but because it does not include color photos one of the other bird guides would be a better model.

1180. Jefferson, Thomas A., Stephen Leatherwood, and Marc A. Webber. **Marine Mammals of the World**. Rome: United Nations Environment Programme, Food and Agriculture Organization of the United Nations, 1993. 320p. (FAO Species Identification Guide). $65.00pa. ISBN 92571032920pa.

    119 species; illustrated keys; black-and-white illustrations of animals and skulls; maps. Species accounts include English, French, and Spanish common names, distinctive characteristics, confusing species, size, distribution, behavior, exploitation, and status.

1181. Jones, J. Knox, Jr., and Richard W. Manning. **Illustrated Key to Skulls of Genera of North American Land Mammals**. Lubbock, TX: Texas Tech University Press, 1992. 75p. $9.95pa. ISBN 0896722899pa.

    Key throughout; black-and-white illustrations and photos.

1182. Knutson, Roger M. **Flattened Fauna: A Field Guide to Common Animals of Roads, Streets, and Highways**. Berkeley, CA: Ten Speed, 1987. 96p. $6.95pa. ISBN 0898151864pa.

    14 groups or species of reptiles and amphibians, 15 birds, 16 mammals; quick key to major groups; silhouettes of animals

that have "lost [their] third dimension"; "death list." Despite the humorous intent of this guide to "road fauna," it does cover most of the species a traveler is likely to see alongside the road (except deer and other large mammals, since they will not have lost their third dimension), and discusses genuine natural history, including why the animals are found by roads. Not as gross as some of the "roadkill cafe" jokes, honest.

1183. Leatherwood, Stephen, and Randall R. Reeves. **The Sierra Club Handbook of Whales and Dolphins**. San Francisco: Sierra Club, 1983. 302p. $18.00pa. ISBN 0871563401pa; 087156341X.

80 species; color illustrations and black-and-white photos; natural history. Illustrated by Larry Foster.

1184. McDougall, Len. **The Complete Tracker: Tracks, Signs, and Habits of North American Wildlife**. New York: Lyons & Burford, 1997. 273p. $14.95pa. ISBN 1558214585pa.

45 species; visual key to tracks; black-and-white photos and illustrations; maps. Tracks and scats are illustrated; extensive natural history.

1185. Miles, Linda, and Betty Wilson. **The Complete Field Guide to Kitty Cat Positions**. Stamford, CT: Longmeadow, 1993. 120p. $6.95pa. ISBN 0681417889pa.

Humorous cartoon guide to cat poses.

1186. Murie, Olaus J. **A Field Guide to Animal Tracks**. 2nd ed. Boston: Houghton Mifflin, 1975. 375p. (Peterson Field Guide Series, No. 9). $18.00pa. ISBN 0395910983pa.

All mammals in North America; key; black-and-white illustrations. Illustrated by the author.

1187. Pukite, John. **A Field Guide to Cows**. New York: Viking Penguin, 1998. 141p. $11.95pa. ISBN 0140273883pa.

52 breeds; key; black-and-white illustrations; life list; descriptions include origin of breeds and purpose. A light-hearted but genuine guide, just the thing to take along on a drive in the country. Originally published in 1996 by Falcon Press. The author has plans to prepare a similar guide for pig watching.

❑ Reader's Digest Editors. **North American Wildlife: Mammals, Reptiles, and Amphibians**. See entry 725.

1188. Reeves, Randall R., Stephen Leatherwood, and Brent Stewart. **The Sierra Club Handbook of Seals and Sirenians**. San Francisco: Sierra Club, 1992. 359p. $18.00pa. ISBN 0871566567pa.

All 42 species worldwide; black-and-white photos and color illustrations; natural history.

1189. Reid, Fiona. **A Field Guide to the Mammals of Central America and Southeast Mexico**. New York: Oxford University Press, 1997. 456p. $55.00; $29.95pa. ISBN 0195064003; 0195064011pa.

400 species; color illustrations; maps; multiple common names in English, Spanish, and local languages; natural history. Also includes a guide to major parks and lists of animals found in each. Illustrated by the author.

1190. Schober, Wilfried, and Eckard Grimmberger. **The Bats of Europe & North America: Knowing Them, Identifying Them, Protecting Them**. Neptune City, NJ: T. F. H., 1997. 239p. $29.95. ISBN 0793804906.

30 species of European bats and 39 North American bats; key to European bats; color and black-and-white photos; natural history and extensive background. More information is presented on the European species than on the American ones.

1191. Searfoss, Glenn. **Skulls and Bones: A Guide to the Skeletal Structures and Behavior of North American Mammals**. Mechanicsburg, PA: Stackpole, 1995. 277p. $19.95pa. ISBN 0811725715pa.

More of a general textbook of comparative skeletal anatomy than a field guide. However, it contains line drawings of the larger bones of common North American mammals, arranged by bone type to aid in the identification of bones found in the field. Also includes information on collecting and preparing specimens. Illustrated by the author.

❑ Smith, Richard P. **Animal Tracks and Signs of North America**. See entry 730.

1192. Stackpole Company. **A Guide to Animal Tracks**. Harrisburg, PA: Stackpole, 1976. 96p. $5.95pa. ISBN 0811722546pa.

43 species of game animals; black-and-white illustrations of tracks and animals; descriptions include general characteristics, range, breeding, and food. Pocket sized.

1193. Sylvestre, Jean-Pierre. **Dolphins & Porpoises: A Worldwide Guide**. New York: Sterling, 1993. 160p. $29.95. ISBN 080698791X.

43 species; color illustrations and photographs; maps; includes skulls of some species.

1194. Whitaker, John O. **The Audubon Society Field Guide to North American Mammals**. Rev. ed. New York: Knopf, 1996. 937p. (Audubon Society Field Guide Series). $19.00 (flexi). ISBN 0679446311 (flexi).

380 species; color photographs; key to tracks; natural history.

1195. Zim, Herbert Spencer, and Donald F. Hoffmeister. **Mammals: A Guide to Familiar American Species: 218 Animals in Full Color**. New York: Golden, 1991. 160p. (Golden Guide). $5.95pa. ISBN 0307640582; 0307240584pa.

Color illustrations.

❑ **Wildlife Identification Pocket Guide: With Field Dressing Commentary**. See entry 732.

Eastern North America

1196. Brown, Larry N. **A Guide to the Mammals of the Southeastern United States**. Knoxville: University of Tennessee Press, 1997. 236p. ISBN 0870499653; 0870499661pa.

132 species; key to orders; black–and–white and color photos; maps; checklist; natural history. An oversized handbook rather than a true field guide.

1197. Chapman, William K., and Dennis Aprill. **Mammals of the Adirondacks**. Utica, NY: North Country, 1991. 159p. $13.95pa. ISBN 0932052614pa.

100 species; keys; color photos separate from text; natural history.

1198. Choate, Jerry R., J. Knox Jones, Jr., and Clyde Jones. **Handbook of Mammals of the South-Central States**. Baton Rouge: Louisiana State University Press, 1994. 304p. $40.00. ISBN 0807118192.

110 species; black–and–white photos; keys; maps; natural history. Manual/field guide; covers Alabama, Arkansas, Georgia, Kentucky, Louisiana, Mississippi, and Tennessee.

1199. Godin, Alfred J. **Wild Mammals of New England: Field Guide Edition**. Yarmouth, ME: DeLorme, 1981. 207p. ISBN 0899330126.

91 species, plus 7 locally extinct; black-and-white illustrations, maps; natural history. Condensed version of author's handbook by the same name. Illustrated by the author.

1200. Grierson, Ruth Gortner. **Acadia National Park: Wildlife Watcher's Guide**. Minocqua, WI: NorthWord, 1995. 94p. $11.95pa. ISBN 1559714557pa.

50 species; color photos. Includes tracks, natural history, and wildlife viewing tips.

❑ Headstrom, Richard. **Identifying Animal Tracks: Mammals, Birds, and Other Animals of the Eastern United States**. See entry 740.

1201. Jones, J. Knox, Jr., and Elmer C. Birney. **Handbook of Mammals of the North-Central States**. Minneapolis: University of Minnesota Press, 1988. 346p. $39.95, $18.95pa. ISBN 0816614199; 0816614202pa.

99 native and 5 introduced species; black-and-white photos.

1202. Jones, J. Knox, Jr., David M. Armstrong, and Jerry R. Choate. **Guide to Mammals of the Plains States**. Lincoln: University of Nebraska Press, 1985. 371p. ISBN 0803275579pa; 0803225628.

135 native and 8 introduced species; key; color photos; maps; handbook. Covers Kansas, Nebraska, North Dakota, Oklahoma, and South Dakota.

1203. Katona, Steven K., Valerie Rough, and David T. Richardson. **A Field Guide to Whales, Porpoises, and Seals: From Cape Cod to Newfoundland**. 4th ed., rev. Washington, DC: Smithsonian Institution Press, 1993. 316p. $15.95pa. ISBN 1560983337pa.

22 whales and 7 marine mammals, plus a few other large sea animals; key to stranded mammals; black-and-white photos and line drawings; extensive natural history.

1204. Kidwell, Al. **Whales and Seals: A Guide to Coastal and Offshore Mammals**. Freeport, ME: DeLorme, 1983. 48p. (A Maine Geographic book.) ISBN 0899330517pa.

19 species of whales and 3 seals; color illustrations; natural history. Covers the Gulf of Maine from Cape Cod to Cape Sable, Nova Scotia. Illustrated by Jon Luoma.

1205. Kurta, Allen. **Mammals of the Great Lakes Region**. Rev. ed. Ann Arbor, MI: University of Michigan Press, 1995. 376p. $44.50; $16.95pa. ISBN 0472094971; 0472064975pa.

Comprehensive; keys; black-and-white photos and illustrations; natural history. Rather technical; includes information on how to study mammals. Revised edition of William Henry Burt's *Mammals of the Great Lakes Region*, which was a revision of Burt's *The Mammals of Michigan*. Illustrated by Scott A. Schwemmin.

1206. Linzey, Donald W. **Mammals of Great Smoky Mountains National Park**. Blacksburg, VA: McDonald & Woodward, 1994. 140p. $14.95pa. ISBN 0939923483pa.

70 species; color photos. Includes natural history and distribution of species within the national park.

1207. Miller, Dorcas S. **Track Finder: A Guide to Mammal Tracks of Eastern North America**. Berkeley, CA: Nature Study Guild, 1981. 60p. $3.00pa. ISBN 0912550120pa.

65 species; key; line drawings; pocket sized. Illustrated by Cherie Hunter Day.

1208. Milstein, Michael. **Badlands, Teddy Roosevelt, and Wind Cave National Parks: Wildlife Watcher's Guide**. Minocqua, WI: NorthWord, 1996. 96p. $11.95pa. ISBN 1559715758pa.

50 species; color photos. Includes tracks, natural history, and wildlife viewing tips.

1209. Schmidly, David J. **The Bats of Texas**. College Station, TX: Texas A&M University Press, 1991. 188p. (W.L. Moody Jr. Natural History Series, No. 11). $34.50; $19.95pa. ISBN 0890964033; 0890964505pa.

32 species; black-and-white photos and illustrations, some color photos; maps; keys; extensive natural history. Illustrated by Christine Stetter.

1210. Sheldon, Ian. **Animal Tracks of Ontario**. Renton, WA: Lone Pine, 1997. 159p. $5.95pa. ISBN 1551051095pa.

Not seen; see series entry 3.

1211. Sheldon, Ian. **Animal Tracks of the Great Lakes Region**. Edmonton: Lone Pine, 1997. 160p. $5.95pa. ISBN 1551051079pa.

60 species, including a few birds and reptiles; black-and-white illustrations; descriptions include natural history and similar species. Includes separate "Identification Chart" comparing track patterns and prints.

1212. Stall, Chris. **Animal Tracks of New England: Connecticut, Maine, Massachusetts, Rhode Island, New Hampshire, and Vermont**. Seattle: Mountaineers, 1989. 112p. $5.95pa. ISBN 0898861942pa.

 45 species, including birds, snakes, and mammals; drawings of tracks only. Pocket sized.

1213. Stall, Chris. **Animal Tracks of Texas**. Seattle: Mountaineers, 1990. 111p. $6.95pa. ISBN 0898862280pa.

 Not seen; see series entry 4.

1214. Stall, Chris. **Animal Tracks of the Great Lakes States: Illinois, Indiana, Michigan, Minnesota, New York, Pennsylvania, Ohio, and Wisconsin**. Seattle: Mountaineers, 1989. 112p. $5.95pa. ISBN 0898861969pa.

 40 mammals, 7 birds; black-and-white illustrations; pocket sized.

1215. Stall, Chris. **Animal Tracks of the Mid-Atlantic Region**. Seattle: Mountaineers, 1989. 112p. $4.95pa. ISBN 0898861977pa.

 45 species; line drawings; pocket sized.

1216. Stall, Chris. **Animal Tracks of the Southeast States**. Seattle: Mountaineers, 1989. 120p. $5.95pa. ISBN 089886223Xpa.

 48 species; black-and-white illustrations of tracks; descriptions include order, family, range and habitat, size, diet, and comments.

1217. Webster, William David, James F. Parnell, and Walter C. Biggs. **Mammals of the Carolinas, Virginia, and Maryland**. Chapel Hill: University of North Carolina Press, 1985. 255p. $24.95. ISBN 0807816639.

 121 species; color photos; natural history.

1218. Whitaker, John O., and William John Hamilton. **Mammals of the Eastern United States**. 3rd ed. Ithaca, NY: Comstock, 1998. 608p. $50.00. ISBN 0801434750.

 121 species; keys; color and black-and-white photos; maps. Species accounts include extensive natural histories. This is an oversized handbook rather than a portable field guide, but is a good source of information on identification and natural history.

Western North America

1219. Alaska Geographic. **Mammals of Alaska: A Comprehensive Guide**. Anchorage, AK: Alaska Geographic Society, 1996. 176p. (Alaska Geographic Guides). $17.95pa. ISBN 1566610346pa.

 75 species; color photos; natural history. Oversized.

1220. Armstrong, David Michael. **Rocky Mountain Mammals: A Handbook of Mammals of Rocky Mountain National Park and Vicinity**. Rev. ed. Boulder: Colorado Associated University Press in cooperation with Rocky Mountain Nature Association, 1987. 223p. ISBN 0870811681, 087081169Xpa.

 65 species; keys; color photos and black-and-white illustrations; checklist; handbook. Illustrated by Bill Border.

1221. Balcomb, Kenneth C., III. **The Whales of Hawaii, Including All Species of Marine Mammals in Hawaiian and Adjacent Water**. San Francisco, CA: Marine Mammal Fund, 1987. 99p. $9.95pa. ISBN 0961780304pa.

 23 species of whales and dolphins, plus the monk seal; color photos; natural history. Illustrated by Larry Foster.

1222. Barwise, Joanne E. **Animal Tracks of Western Canada**. Edmonton: Lone Pine, 1989. 127p. $6.95pa. ISBN 0919433200pa.

 36 common species; key; black-and-white illustrations of tracks and mammals; maps. Pocket sized. Illustrated by Ozlem Boyacioglu.

1223. Clark, Tim W., and Mark R. Stromberg. **Mammals in Wyoming**. Lawrence, KS: University of Kansas, Museum of Natural History, 1987. 314p. (Public Education Series, No. 10). $25.00; $12.95pa. ISBN 0893380261; 0893380253pa.

 117 species; black-and-white photos; maps; natural history.

1224. Cockrum, E. Lendell. **Mammals of the Southwest**. Tucson, AZ: University of Arizona Press, 1982. 176p. ISBN 0816507600; 0816507597pa.

 171 species, 80 illustrated; black-and-white drawings; maps. Illustrated by Sandy Truett.

1225. Flaherty, Chuck. **Whales of the Northwest: A Guide to the Marine Mammals of Oregon, Washington and British Columbia**. Seattle, WA: Cherry Lane, 1990. 23p. $5.95pa. ISBN 0962675008pa.

 17 species; color photos; natural history; pamphlet.

☐ Halfpenny, James C. **Scats and Tracks of the Rocky Mountains**. See entry 747.

1226. Hoffmeister, Donald Frederick. **Mammals of the Grand Canyon**. Urbana: University of Illinois Press, 1971. 183p. ISBN 0252001540; 0252001559pa.

74 species; keys; black-and-white illustrations; natural history. Illustrated by James Gordon Irving.

1227. Illg, Gordon. **Rocky Mountain Safari: A Wildlife Discovery Guide**. Niwot, CO: Roberts Rinehart, 1994. 88p. ISBN 1879373793pa.

37 species; color photos; natural history and where to locate wildlife.

1228. Jameson, E. W., and Hans J. Peeters. **California Mammals**. Berkeley, CA: University of California Press, 1986. 403p. (California Natural History Guides, No. 52). ISBN 0520052528; 0520053915pa.

Comprehensive; keys; black-and-white and color photos; maps; includes illustrations of skulls.

1229. Kaill, Michael. **Field Guide to Alaskan Whales**. Anchorage, AK: Anchorage Audubon Society, 1984. 28p.

13 species; black-and-white illustrations; descriptions include appearance and behavior of whales at the surface. Has tips on whale watching, including sites in Alaska.

1230. Kreitman, Richard C., and Mary Jane Schramm. **West Coast Whale Watching: The Complete Guide to Observing Marine Mammals**. San Francisco: HarperCollinsWest, 1995. $15.00pa. ISBN 006258619Xpa.

29 species of whales and dolphins, 6 species of seals and sea otters; black-and-white illustrations. This is a guidebook, not a field guide.

1231. Larrison, Earl Junior. **Mammals of the Northwest: Washington, Oregon, Idaho, and British Columbia**. Seattle: Seattle Audubon Society, 1976. 256p. (Trailside Series). ISBN 0914516043.

187 species; black-and-white photos and illustrations; checklist; natural history. Includes marine mammals. Published in 1970 as *Washington Mammals*. Illustrated by Amy C. Fisher.

1232. Larrison, Earl Junior, and Donald R. Johnson. **Mammals of Idaho**. Moscow, ID: University Press of Idaho, 1981. 166p. (Northwest Naturalist Books). ISBN 0893010707pa.

110 species; key to skulls; black-and-white illustrations (not all species are illustrated); natural history. Revised edition of the author's 1967 *Guide to Idaho Mammals*. Illustrated by Gregory A. Pole.

1233. Mays, Buddy. **Guide to Western Wildlife**. Rev. 3rd ed. San Francisco, CA: Chronicle, 1988. 221p. $8.95pa. ISBN 087701504X.

50 species; black-and-white photos; includes drawings of tracks and skulls; natural history. Arranged by size, includes apparently serious coverage of Bigfoot!

1234. Olin, George. **Mammals of the Southwest Deserts**. Rev ed. Globe, AZ: Southwest Parks and Monuments Association, 1975. 102p. (Southwestern Monuments Association. Popular Series, No. 8). ISBN 0911408363.

50 species; color illustrations; natural history. More of an attractive tourist guide than a working field guide. Illustrated by Jerry Cannon.

1235. Olin, George. **Mammals of the Southwest Mountains and Mesas**. Globe, AZ: Southwestern Monuments Association, 1961. 126p. (Southwestern Monuments Association. Popular Series, No. 9).

42 species; black-and-white illustrations; maps; primarily a natural history. Illustrated by Edward Bierly.

1236. Orr, Robert Thomas, and Roger C. Helm. **Marine Mammals of California**. Rev. ed. Berkeley, CA: University of California Press, 1989. 93p. (California Natural History Guides, No. 29). $11.95pa; $39.95. ISBN 0520065158pa; 0520065352.

32 species; black-and-white illustrations, some photos; natural history; sites for viewing. Illustrated by Jacqueline Schonewald.

1237. Osborne, Richard, John Calambokidis, and Eleanor M. Dorsey. **A Guide to Marine Mammals of Greater Puget Sound**. Anacortes, WA: Island, 1988. 191p. ISBN 0961558016pa.

24 species of cetaceans, pinnipeds, and otters; black-and-white and some color photos; maps. Descriptions include identification, natural history, distribution in Puget Sound, and how to observe. The front endpage features illustrations of whales as seen at the surface of the ocean, and the back endpage does the same for the pinnipeds. The book also includes identification marks of the 82 orcas and 28 minke whales that were residents in Puget Sound at the time.

1238. Pandell, Karen, and Chris Stall. **Animal Tracks of the Pacific Northwest**. Seattle: Mountaineers, 1981. 112p. $6.95pa. ISBN 0898860121pa.

40 mammals, 9 birds; life-sized footprints. Pocket sized.

1239. Russo, Ron. **Mountain State Mammals**. Berkeley, CA: Nature Study Guild, 1991. 132p. $5.25pa. ISBN 091255021Xpa.

100 species; key; black-and-white illustrations, including skulls, tracks, scats; natural history; pocket sized. Illustrated by Barbara Downs.

1240. Russo, Ron. **Pacific Coast Mammals: A Guide to Mammals of the Pacific Coast States, Their Tracks, Skulls, and Other Signs**. Berkeley, CA: Nature Study Guild, 1987. 92p. $4.25pa. ISBN 0912550163pa.

50 species; keys; line drawings; includes skulls, tracks, and signs. Formerly published as *Mammal Finder*.

1241. Sheldon, Ian. **Animal Tracks of Northern California**. Edmonton: Lone Pine, 1997. $5.95pa. ISBN 1551051036pa.

Not seen; see series entry 3.

1242. Sheldon, Ian. **Animal Tracks of Southern California**. Edmonton: Lone Pine, 1998. $5.95pa. ISBN 1551051052pa.

Not seen; see series entry 3.

1243. Sheldon, Ian. **Animal Tracks of the Rockies**. Edmonton: Lone Pine, 1997. 159p. $5.95pa. ISBN 1551050897pa.

54 mammals, 7 birds and herps; black-and-white illustrations of animal and tracks; descriptions include natural history and similar species. Includes separate "Identification Chart" comparing track patterns and prints.

1244. Sheldon, Ian. **Animal Tracks of Washington and Oregon**. Redmond, WA: Lone Pine, 1997. 159p. $5.95pa. ISBN 1551050900pa.

Not seen; see series entry 3.

1245. Shirley, Gayle Corbett. **Montana Wildlife: A Children's Field Guide to the State's Most Remarkable Animals**. Billings, MT: Falcon, 1993. 48p. (Interpreting the Great Outdoors). $7.95pa. ISBN 1560441542pa.

75 species of mammals; color illustrations; arranged by habitat; includes multiple common names. Illustrated by Sandy Allnock.

1246.   Simons, L. L. **Hot on the Trail: The Only Alaska Wildlife Guide You'll Ever Need**. Juneau, AK: Cold Country, 1993. 69p. $6.75pa. ISBN 0963487701pa.

   12 species; black-and-white illustration of mammals, color photos of scats; natural history and notes. Includes an identification guide to scats found along trails, since scats are more likely to be seen than the animals that made them.

1247.   Smith, Dave. **Alaska's Mammals: A Guide to Selected Species**. Anchorage: Alaska Northwest, 1995. 94p. (Alaska Pocket Guide). $12.95pa. ISBN 0882404636pa.

   22 land and 13 marine species; color photos; background information and natural history.

1248.   Stall, Chris. **Animal Tracks of Alaska**. Seattle, WA: Mountaineers, 1993. 112p. $5.95pa. ISBN 089886352Xpa.

   35 mammals and 10 birds; black-and-white illustrations; pocket sized.

1249.   Stall, Chris. **Animal Tracks of Northern California**. Seattle: Mountaineers, 1989. 120p. $6.95pa. ISBN 0898861950pa.

   Not seen; see series entry 4.

1250.   Stall, Chris. **Animal Tracks of Southern California**. Seattle: Mountaineers, 1990. 124p. $5.95pa. ISBN 089886237Xpa.

   Not seen; see series entry 4.

1251.   Stall, Chris. **Animal Tracks of the Rocky Mountains: Idaho, Montana, Wyoming, Utah, Colorado, Arizona, and New Mexico**. Seattle: Mountaineers, 1989. 112p. $5.95pa. ISBN 0898861853pa.

   42 mammals and 4 birds; line drawings; pocket sized.

1252.   Stall, Chris. **Animal Tracks of the Southwest States: Arizona, New Mexico, Southern Utah, and Southwest Colorado**. Seattle: Mountaineers, 1990. 112p. $5.95pa. ISBN 0898862264pa.

   45 species; black-and-white illustrations; pocket sized.

1253.   Ulrich, Tom J. **Mammals of the Northern Rockies**. Missoula, MT: Mountain, 1986. 157p. $12.00pa. ISBN 0878422005pa.

   58 species; color photos; natural history; arranged with large mammals first. Covers the area from the Canadian Rockies to the Tetons.

1254.   Wassink, Jan L. **Mammals of the Central Rockies**. Missoula, MT: Mountain, 1993. 161p. $14.00pa. ISBN 0878422374pa.

> 50 species; color photos; natural history. Covers Colorado, Wyoming, Utah, and southeast Idaho.

1255.   Wile, Darwin. **Identifying and Finding the Mammals of Jackson Hole (Including Grand Teton National Park): A Field Guide**. Jackson, WY: D. Wile, 1996. 139p. $9.95pa.

> 60 species; black-and-white and color photos; descriptions include scats and tracks, status, natural history, and where to find mammals.

1256.   Wynne, Kate. **Guide to Marine Mammals of Alaska**. Fairbanks, AK: Alaska Sea Grant College Program, University of Alaska, Fairbanks, 1992. 75p. (MAB Series, Alaska Sea Grant College Program, No. 44). $17.95pa. ISBN 1566120098pa.

> 30 species; color photos; maps; includes natural history and silhouettes of similar species; common names in English, Japanese, Russian, and one or more native languages (Aleut, Yupik, or Inupiat). Spiralbound.

1257.   Zeveloff, Samuel I. **Mammals of the Intermountain West**. Salt Lake City: University of Utah Press, 1988. 365p. $14.95pa. ISBN 0874802962, 0874803276pa.

> 151 species; color and black-and-white illustrations; maps; checklist; natural history; covers area between Rocky Mountains and Sierra Nevadas, including all of Utah, most of Nevada, and parts of California, Oregon, Idaho, and Wyoming. Illustrated by Farrell A. Collett.

Astronomy and the Weather

The uniting feature of the guides in this chapter is their emphasis on objects and/or phenomena found above our heads. There are many guides to astronomy and the weather that are not included here because they do not fit the field guide definition, either because they do not emphasize identification or because they are not appropriate for outdoor or field use. However, almost all of the usual field guide series include these topics, and they are certainly of interest to amateur naturalists.

Unlike the other chapters, this one is not subdivided by region, because most of the guides cover a wide area. All of the astronomy guides included here are for the Northern Hemisphere, unless otherwise indicated. Weather guides generally emphasize the weather of their country of origin, but most include weather phenomena that occur in other parts of the world or which have wide impact, such as tornadoes or monsoons. Some of the comprehensive guides in Chapter 2, "Flora and Fauna," also include information on astronomy and the weather.

1258. Baker, David. **The Henry Holt Guide to Astronomy**. 1st American ed. New York: Holt, 1990. 288p. ISBN 0805011978pa.

Color photos and illustrations, black-and-white illustrations; has Northern and Southern Hemisphere constellation charts, plus charts of individual constellations; includes background information. Also published as the *Hamlyn Guide to Astronomy* and the *Larousse Guide to Astronomy*. Illustrated by David A. Hardy.

1259. **Cambridge Pocket Star Finder: A Month-by-Month Guide to the Night Sky**. New York: Cambridge University Press, 1996. 50p. $6.95pa. ISBN 0521589932pa.

12 monthly sky charts for the Northern Hemisphere. First published in 1918 as *Stars at a Glance*, revised and retitled *Star Finder*.

1260. Chartrand, Mark R. **The Audubon Society Field Guide to the Night Sky**. New York: Knopf, 1991. 714p. (Audubon Society Field Guide Series). $19.00. ISBN 0679408525.

48 monthly charts, 88 constellation maps; includes color photos of constellations, meteors, etc.

1261. Chartrand, Mark R. **Planets: A Guide to the Solar System**. New York: Golden, 1990. 160p. $5.50pa. ISBN 0307640779; 0307240770pa.

Contains color illustrations and photos of the planets, comets, eclipses, and other objects within our solar system; also includes star charts for locating planets.

1262. Chartrand, Mark R. **Skyguide: A Field Guide for Amateur Astronomers**. New York: Golden, 1982. 280p. (Golden Field Guide Series). $11.95pa. ISBN 0307136671pa; 0307470105.

35 seasonal sky charts for the Northern Hemisphere and 180 constellation maps; solar system maps. Illustrated by Helmut K. Wimmer.

1263. Cox, John. **Cambridge Pocket Star Atlas**. New York: Cambridge University Press, 1996. 48p. $6.95pa. ISBN 0521589924pa.

Pocket-sized atlas with 14 maps and brief notes for naked-eye viewing; little text. Tables with information on planets, eclipses, meteor showers, stars, deep-sky objects. Originally published in the U.K. as *Philip's Pocket Star Atlas*.

1264. Day, John A., and Vincent J. Schaefer. **Peterson First Guide to Clouds and Weather**. Boston: Houghton Mifflin, 1991. 128p. $5.95pa. ISBN 0395906636pa.

Color photos of clouds and other weather-related phenomena.

1265. Dunlop, Storm. **Astronomy: A Step-by-Step Guide to the Night Sky**. New York: Collier, 1985. 192p. (Macmillan Field Guides). ISBN 0020796501pa.

Contains 8 constellation and 8 star charts, plus numerous illustrations of stars, planets, auroras, and other phenomena.

1266. Dunlop, Storm, and Francis Wilson. **Weather and Forecasting**. 1st Collier Books ed. New York: Collier, 1987. 160p. (Macmillan Field Guides). $10.95pa. ISBN 0020137001pa.

Covers weather phenomena and forecasting; includes maps, color and black-and-white photos; worldwide. Previously published as *The Larousse Guide to Weather Forecasting*.

1267. Ekrutt, Joachim. **Stars**. 2nd ed. Hauppauge, NY: Barron's Educational Series, 1996. 80p. (Mini Fact Finders). $5.95pa. ISBN 0812047788pa.

Monthly star charts with events through 1999. Translation of *Sterne*.

1268. Ekrutt, Joachim W. **Stars and Planets: Identifying Them, Learning About Them, Experiencing Them, With All Important Celestial Events up to the Year 2000**. Hauppage, NY: Barron's Educational Series, 1992. 159p. $12.95pa. ISBN 0812047761pa.

175 maps and illustrations; 30 color photos. Translation of *Sterne und Planeten*.

1269. Forey, Pamela, and Cecilia Fitzsimons. **An Instant Guide to Stars and Planets: The Sky at Night Described and Illustrated in Color**. New York: Bonanza, 1988. 124p. ISBN 0517635496.

Includes 50 star and constellation charts, plus numerous illustrations of the solar system; checklist/index.

1270. Henbest, Nigel. **Spotter's Guide to the Night Sky**. 1st American ed. New York: Mayflower, 1979. 64p. ISBN 0831763760pa, 0831763752.

Northern and Southern Hemisphere star charts, 88 constellations, plus information on stars, the solar system, meteorites and night-time atmospheric phenomena. Includes scorecard for recording sightings. Also published in Spanish by Juventud.

1271. Holle, Ronald L., and Richard A. Keen. **Clouds and Storms**. New York: Knopf, 1994. 192p. (National Audubon Society Pocket Guides). $8.00pa. ISBN 067977999Xpa.

80 color photos of weather phenomena.

1272. Kahl, Jonathan D. **National Audubon Society First Field Guide to the Weather**. New York: Scholastic, 1998. 160p. $17.95; $10.95pa. ISBN 0590054694; 0590054880pa.

About a third of the book consists of general information about the weather, while the remainder consists of color photos of cloud formations and other weather phenomena. Includes spotter's chart with small illustrations of weather types.

1273. Karkoschka, Erich. **The Observer's Sky Atlas: With 50 Star Charts Covering the Entire Sky**. New York: Springer-Verlag, 1990. 30p. ISBN 0387515887pa.

Star charts of all stars visible to the naked eye, plus enlarged sections for binocular observation of interesting sky objects;

Northern and Southern Hemispheres. Translation of *Atlas für Himmelsbeobachter*.

1274.   Kerrod, Robin. **The Star Guide: Learn How to Read the Night Sky Star by Star**. 1st Prentice Hall ed. New York: Prentice Hall General Reference, 1993. 160p. $27.50. ISBN 0671874675.

Color photos and illustrations; maps of Northern and Southern Hemisphere skies by season and month; 15 key constellations; photos of planets, including moon maps; background information at end. Includes planisphere.

1275.   Lehr, Paul E., R. Will Burnett, Herbert S. Zim. **Weather: Air Masses, Clouds, Rainfall, Storms, Weather Maps, Climate**. Rev. ed. New York: Golden, 1987. 160p. (Golden Guide). $6.00pa. ISBN 0307640515; 0307240517pa.

A general introduction to weather, rather than an aid to predicting or identifying weather phenomena; color illustrations. Illustrated by Harry McNaught.

1276.   Leventer, Amy, and Geoffrey O. Seltzer. **Earth from Space**. New York: Knopf, 1995. 192p. (National Audubon Society Pocket Guides). $8.00pa. ISBN 0679760571pa.

80 images, mostly color photos of earth taken from space.

1277.   Ludlum, David M. **The Audubon Society Field Guide to North American Weather**. New York: Knopf, 1991. 656p. (Audubon Society Field Guide Series). $19.00 (flexi). ISBN 0679408517 (flexi).

378 color photos of weather and other phenomena; line drawings; background information.

1278.   Mayall, R. Newton, Margaret Mayall, and Jerome Wyckof. **Sky Observer's Guide**. Rev. ed. New York: Golden, 1985. 160p. $5.50pa. ISBN 0307240096pa.

Charts, maps, etc.; primarily how to observe the sky.

1279.   Mechler, Gary. **Constellations of the Northern Sky**. New York: Knopf, 1994. 192p. (National Audubon Society Pocket Guides). $8.00pa. ISBN 0679779981pa.

Sky maps of 54 constellations visible in the Northern Hemisphere.

1280.   Mechler, Gary. **Galaxies and Other Deep-Sky Objects**. New York: Knopf, 1995. 192p. (National Audubon Society Pocket Guides). $9.00pa. ISBN 0679779965pa.

80 color photos of objects visible to amateur sky-watchers.

1281. Mechler, Gary, and Steven Kent Croft. **Planets and Their Moons**. New York: Knopf, 1994. 192p. (National Audubon Society Pocket Guides). $8.00pa. ISBN 0679779973pa.

    80 color photos of planets and their moons, asteroids, and comets.

1282. Mechler, Gary, Robert Marcialis, and Melinda Hutson. **Sun and the Moon**. New York: Knopf, 1994. 192p. (National Audubon Society Pocket Guides). $8.00pa. ISBN 0679760563pa.

    80 color photos of the sun and moon; moon maps.

1283. Number not used.

1284. Moore, Patrick, and Wil Tirion. **Cambridge Guide to Stars and Planets**. New York: Cambridge University Press, 1997. 256p. $14.95pa. ISBN 0521585821pa.

    Charts of 88 constellations of Northern and Southern Hemispheres; color photos; includes background information, chapters on planets, other sky objects. Previously published in the U.K. as *Philip's Guide to Stars and Planets*.

1285. Pasachoff, Jay M. **Field Guide to the Stars and Planets**. 3rd ed. Boston: Houghton Mifflin, 1992. 502p. (Peterson Field Guide Series, No. 15). $18.00pa; $27.00. ISBN 0395911001; 0395910994pa.

    72 sky maps, 52 atlas charts, plus moon maps, timetables, etc.; color and black-and-white photos. Revised edition of Donald H. Menzel's *A Field Guide to the Stars and Planets*. Illustrated by Wil Tirion.

1286. Pasachoff, Jay M. **Peterson First Guide to Astronomy**. Boston: Houghton Mifflin, 1986. 120p. $5.95pa. ISBN 0395935423pa.

    12 sky maps, plus 11 constellation charts; information on planets, galaxies, etc. Illustrated by Wil Tirion and Robin Brickman.

1287. Pasachoff, Jay M. **Peterson First Guide to the Solar System**. Boston: Houghton Mifflin, 1990. 128p. $4.95pa. ISBN 0395524512pa.

    100 color photos, planet finding maps. Illustrated by Wil Tirion.

1288.   Ridpath, Ian. **Astronomy**. New York: Gallery, 1990. 192p. (American Nature Guides). ISBN 0831769696pa.

      22 star charts, 8 moon maps; color illustrations and photos.

1289.   Ridpath, Ian, and Wil Tirion. **Collins Pocket Guide to Stars and Planets**. 2nd ed. London: HarperCollins, 1993. 384p. ISBN 0002199793pa.

      100 sky charts for Northern and Southern Hemispheres, 88 constellation charts, moon maps; color photos; information on choosing a telescope, planets, etc. First edition also published in the U.S. as *Universe Guide to Stars and Planets*.

1290.   Roberts, Charles W., and C. E. N. Frankcom. **Maritime Meteorology: A Guide for Deck Officers**. New York: W. S. Heinman, 1985. 258p. ISBN 0900335912.

      One of the more unusual guides; a study manual and reference for merchant marine deck officers. Includes black-and-white photos and meteorological maps.

1291.   Ronan, Colin, and Storm Dunlop, eds. **The Skywatcher's Handbook: Night and Day, What to Look for in the Heavens Above**. New York: Crown, 1989. 224p. $16.00pa. ISBN 0517573261pa.

      Color photos and illustrations; daytime phenomena as well as astronomy; extensive background.

1292.   Schaefer, Vincent J., and John A. Day. **A Field Guide to the Atmosphere**. Boston: Houghton Mifflin, 1981. 359p. (Peterson Field Guide Series, No. 26). $21.95; $19.00pa. ISBN 0395240808; 0395330335pa.

      358 photos, mostly black-and-white. Illustrated by Christy E. Day.

1293.   United States. National Weather Service. **Guide to Sea State, Wind, and Clouds**. Washington, DC: U.S. Dept. of Commerce, National Oceanic and Atmospheric Administration, National Weather Service, 1995. 16, 28, 3 leaves.

      Photographs of the sea surface for use in estimating wind speed, plus cloud photos for forecasting weather; for mariners, spiralbound with laminated pages.

1294.   Whitney, Charles Allen. **Whitney's Star Finder: A Field Guide to the Heavens**. 5th ed. New York: Knopf, 1989. 111p. ISBN 0679725822pa.

Book contains extensive background information; has star Locator Wheel in pocket.

1295.   Zim, Herbert Spencer, and Robert H. Baker. **Stars: A Guide to the Constellations, Sun, Moon, Planets, and Other Features of the Heavens**. Rev. ed. New York: Golden, 1990. 160p. (Golden Guide). $5.50pa. ISBN 0307244938pa.

26 constellation maps, 4 seasonal sky charts plus north and south circumpolar maps; information on stars, planets, comets, meteors, auroras, etc. Illustrated by James G. Irving.

Chapter 17

Geology and Fossils

Whereas the previous chapter dealt with phenomena overhead, this one focuses on matters underfoot. These field guides cover rocks, minerals, gemstones, geological formations, and fossils. Worldwide guides are more prevalent in this subject area, since rocks change little from one part of the world to another.

Again, like the field guides in the chapter titled "Weather," these field guides tend to come from the standard field guide series. There are a large number of identification guides to rocks and minerals that do not fit the standard field guide mold due to size or layout. The very numerous geological field guidebooks are also excluded. These field guides, often with titles such as *A Field Guide to the Geology of . . .* or *Field Trip Guide to . . .*, are actually road logs of a field trip.

NORTH AMERICA

1296. Adams, George F., and Jerome Wycoff. **Landforms**. New York: Golden, 1971. 160p. (A Golden Science Guide). ISBN 0307243508pa.
      100 types of landforms; color and black-and-white photos and black-and-white illustrations; arranged by type of landform.

1297. Arduini, Paolo, and Giorgio Teruzzi. **Simon and Schuster's Guide to Fossils**. New York: Simon and Schuster, 1986. 317p. $22.95; $14.00pa. ISBN 0671632191; 0671631322pa.
      260 types; color photos; includes classification, description, stratigraphic position, notes. Translation of *Fossili*.

1298.   Arem, Joel E. **Gems and Jewelry: All Color Guide**. 2nd ed. Phoenix, AZ: Geoscience, 1992. 159p. $9.95pa. ISBN 0945005091pa.

   30 types of gemstones including man-made stones; color photos; background.

1299.   Arem, Joel E. **Rocks and Minerals: All Color Guide**. Phoenix, AZ: Geoscience, 1991. 145p. $8.95pa. ISBN 0945005067pa.

   150 types; color photos; extensive information on general geology.

1300.   Bauer, Jaroslav. **A Field Guide in Color to Minerals, Rocks and Precious Stones**. Secaucus, NJ: Treasure, 1992. 208p. $9.98. ISBN 1555210937.

   550 types; color photos; arranged by color.

1301.   Bauer, Jaroslav, and Vladimir Bouska. **A Guide in Colour to Precious and Semiprecious Stones**. Secaucus, NJ: Chartwell Books, 1989. 240p. $9.98. ISBN 1555213626.

   70 types; color photos of natural and cut stones; extensive background information. Includes maps of major gem mines.

1302.   Bell, Pat, and David Wright. **Rocks and Minerals**. New York: Collier Books, 1985. 192p. (Macmillan Field Guides). ISBN 0020796404pa.

   400 types; key; color photos; extensive introductory material.

1303.   Brocardo, G. **Minerals and Gemstones of the World**. Rev. English ed. Newton Abbot: David and Charles, 1994. 215p. (Naturetrek Guide). $14.95pa. ISBN 0715301977pa.

   150 types; color photos; pictographic tables; arranged by color. Previous edition published as *Minerals and Gemstones: An Identification Guide*.

1304.   Chesterman, Charles Wesley. **The Audubon Society Field Guide to North American Rocks and Minerals**. New York: Knopf, 1978. 850p. (Audubon Society Field Guide Series). $19.00 (flexi). ISBN 0394502698 (flexi).

   780 types; color photos; arranged by color.

1305.   Chesterman, Charles Wesley. **Familiar Rocks and Minerals**. New York: Knopf, 1988. 192p. (Audubon Society Pocket Guides). $9.00pa. ISBN 0394757947pa.

   68 minerals, 12 rocks; color photos.

1306. Cipriani, Curzio, and Alessandro Borelli. **Simon and Schuster's Guide to Gems and Precious Stones**. New York: Simon and Schuster, 1986. 384p. $14.00pa. ISBN 0671601164; 0671604309pa.

    100 varieties; color photos. Translation of *Il Tutto-Pietre Preziose*.

1307. Fejer, Eva, and Cecilia Fitzsimons. **An Instant Guide to Rocks and Minerals: The Most Familiar Rocks and Minerals of North America Described and Illustrated in Full Color**. New York: Bonanza, 1988. 125p. (Instant Guide Series). $4.95. ISBN 051763550X.

    100 types; color illustrations; checklist.

1308. Hall, Cally. **Gemstones**. New York: Dorling-Kindersley, 1994. 160p. (Eyewitness Handbooks). $17.95pa; $29.95. ISBN 1564584984pa; 1564584992.

    130 types of gems; color photos of rough and worked stones.

1309. Hamilton, William Roger, A. R. Woolley, and A. C. Bishop. **The Henry Holt Guide to Minerals, Rocks, and Fossils**. New York: H. Holt, 1989. 320p. $12.95pa. ISBN 0805011188pa.

    500 rock types plus 50 fossils; color photos; extensive introductory material. Previously published as *The Larousse Guide to Minerals, Rocks, and Fossils* and *The Hamlyn Guide to Minerals, Rocks, and Fossils*.

1310. Headstrom, Richard. **Suburban Geology: An Introduction to the Common Rocks and Minerals of Your Back Yard and Local Park**. Englewood Cliffs, NJ: Prentice-Hall, 1985. 136p. ISBN 0138592322pa; 0138592403.

    110 types of minerals, 40 of rocks; black-and-white illustrations and photos; color key with name, hardness, streak color, cleavage for each mineral; separate description section.

1311. Hochleitner, Rupert. **Minerals**. Hauppauge, NY: Barron's Educational Series, 1990. 62p. (Barron's Mini Fact Finders). $4.95pa. ISBN 0812044568pa.

    55 types; color photos; arranged in order of increasing hardness. Descriptions include streak color, distribution, hardness, and chemical formula. Translation of *Mineralien*.

1312. Hochleitner, Rupert. **Minerals: Identifying, Learning About, and Collecting the Most Beautiful Minerals and Crystals**. Hauppauge, NY: Barron's Educational Series, 1994. 237p. (Barron's Nature Guide). $14.95pa. ISBN 0812017773pa.

    Covers 250 varieties; color photos; arranged by color streak. Translation of *Mineralien*.

1313. Hochleitner, Rupert. **Gemstones: Precious Stones, Semi-Precious Stones, Natural Crystal Formations . . .: A Handbook for Amateur Gemologists**. New York: Barron's, 1990. 63p. (Mini Fact Finders). $4.95pa. ISBN 0812044541pa.

    50 types; color photos; pocket sized. Translation of *Edelsteine*.

1314. Horenstein, Sidney. **Familiar Fossils**. New York: Knopf, 1988. 192p. (Audubon Society Pocket Guides). $9.00pa. ISBN 0394757912pa.

    80 species; color photos.

1315. Mayr, Helmut. **A Guide to Fossils**. Princeton, NJ: Princeton University Press, 1992. 256p. $39.50; $19.95pa. ISBN 069108789X; 0691029229pa.

    180 genera covered with 500 excellent photos of prepared specimens. Little introductory material. Translation of *Fossilien*. Also published in Britain by HarperCollins.

1316. Moody, Richard. **Fossils**. New York: Collier Books, 1986. 192p. (Macmillan Field Guides). $11.95pa. ISBN 002063370Xpa.

    Over 300 genera of fossils; color illustrations.

1317. O'Donoghue, Michael. **Rocks and Minerals**. New York: Gallery Books, 1990. 224p. (American Nature Guides). $9.98. ISBN 0831769645.

    300 types; color photos; introductory material. Arranged alphabetically by rock or mineral name.

1318. Pellant, Chris. **Rocks and Minerals**. 1st American ed. New York: Dorling Kindersley, 1992. 256p. (Eyewitness Handbooks). $17.95 (flexi); $30.00. ISBN 156458061X (flexi); 1564580334.

    500 types; color photos; includes descriptions, tests, forms.

1319. Pough, Frederick H. **A Field Guide to Rocks and Minerals**. 5th ed. Boston: Houghton Mifflin, 1995. 480p. (Peterson Field Guide Series, No. 7). $18.00pa. ISBN 0395727782; 039591096Xpa.

    270 types; black-and-white and color photos. Includes information on collecting and finding rocks, plus properties, classification, and description.

1320.   Pough, Frederick H. **Peterson First Guide to Rocks and Minerals**. Boston: Houghton Mifflin, 1991. $5.95pa. ISBN 0395562759pa.

200 types; color photos.

1321.   Prinz, Martin, George Harlow, and Joseph Peters, eds. **Simon and Schuster's Guide to Rocks and Minerals**. New York: Simon and Schuster, 1978. 607p. (A Fireside Book). $15.00pa. ISBN 0671243969; 0671244175pa.

276 minerals and 101 rocks; color photos. Translation of *Minerali e Rocce*, by A. Mottana, R. Crespi, and G. Liborio. Also published in Spanish by Grijalbo.

1322.   Rhodes, Frank Harold Trevor. **Geology**. Rev. ed. New York: Golden, 1991. 160p. (Golden Science Guide). ISBN 0307243494pa.

Covers about 75 topics in geology, such as mass wasting or glaciation. Illustrated by Raymond Perlman.

1323.   Ricciuti, Edward R. **National Audubon Society First Field Guide to Rocks and Minerals**. New York: Scholastic, 1998. 160p. $17.95; $10.95 (flexi). ISBN 0590054635; 0590054848 (flexi).

50 common types covered in detail, another 125 illustrated and described; color photos; extensive introductory material; children's guide. Includes a separate pocket-sized card illustrating the 50 main types of rocks.

1324.   Roberts, David C. **A Field Guide to Geology**. Boston: Houghton Mifflin, 1996. 416p. (Peterson Field Guide Series, No. 47). $27.95; $17.95pa. ISBN 0395663261; 0395663253pa.

Covers the roadside geology of eastern North America, roughly east of the Great Plains. One chapter covers background material; the remainder cover geological formations seen from several major highways in the Canadian Shield, Stable Interior, Appalachian Province, and Coastal Plain regions. Color photos, black-and-white illustrations; maps. A fairly typical example of the geological field guides that are generally excluded from inclusion in this guide. Illustrated by W. Grant Hudson.

1325.   Roberts, John L. **The Macmillan Field Guide to Geological Structures**. London: Macmillan, 1989. 250p. $29.50pa. ISBN 0333421493 (cased); 0333662954pa.

Numerous color photos of structural features that are displayed in rocks in the field such as columnar joints, and cross bedding.

1326.   Schumann, Walter. **Gemstones of the World**. Rev. and expanded ed. New York: Sterling, 1997. 271p. $24.95. ISBN 0806994614.

    75 major types and 300 lesser-known gemstones; color photos of raw and cut stones, including famous gems; extensive introduction and background information; table of constants. Translation of *Edelsteine und Schmucksteine*.

1327.   Schumann, Walter. **Handbook of Rocks, Minerals, and Gemstones**. Boston: Houghton Mifflin, 1993. 380p. $21.00pa. ISBN 0395511372pa; 0395511380.

    Covers 650 varieties; color photos. Published in the U.K. as *Collins Photo Guide to Rocks, Minerals and Gemstones*. Translation of *Der Neue BLV Steine-und Mineralienführer*.

1328.   Schumann, Walter. **Stones and Minerals of the World**. New York: Sterling, 1998. 224p. $14.95pa. ISBN 0806985712pa.

    500 types; color photos; color coded; includes scratch test and other identification tests. Translation of *Mineralien aus aller Welt*.

1329.   Sorrell, Charles A. **Rocks and Minerals: A Field Guide and Introduction to the Geology and Chemistry of Minerals**. New York: Golden, 1973. 280p. (Golden Field Guide Series). $11.95pa. ISBN 0307470059; 0307136612pa.

    500 types; color illustrations; arranged by type of rock or mineral. Original title *Minerals of the World*. Illustrated by George F. Sandstrom.

1330.   Thompson, Ida. **The Audubon Society Field Guide to North American Fossils**. New York: Knopf, 1982. 846p. (Audubon Society Field Guide Series). $19.00 (flexi). ISBN 0394524128 (flexi).

    420 types covered in detail, another 300 briefly described; color photos and black-and-white illustrations; maps.

1331.   Woolley, Alan Robert. **Spotter's Guide to Rocks and Minerals**. 1st American ed. New York: Mayflower Books, 1979. 64p. $12.95; $4.95pa. ISBN 0881109835; 0860201120pa.

    85 types of minerals plus a few fossils; color photos; checklist.

1332.   Zim, Herbert Spencer, and Paul R. Shaffer. **Rocks and Minerals: A Guide to Familiar Minerals, Gems, Ores, and Rocks**. New York: Golden, 1957. 160p. (A Golden Nature Guide). $6.00pa. ISBN 0307635023; 0307244997pa.

    360 types; color illustrations; brief descriptions and background information. Illustrated by Raymond Perlman.

Eastern North America

☐ Beatte, Brian, and Brett Dufur. **River Valley Companion: A Nature Guide**. See entry 123.

☐ Daniel, Glenda. **Dune Country: A Hiker's Guide to the Indiana Dunes**. See entry 129.

1333. Finsley, Charles. **A Field Guide to Fossils of Texas**. 2nd ed. Houston, TX: Gulf, 1996. 211p. (Texas Monthly Field Guide Series). $18.95pa. ISBN 0877192936pa.

Black-and-white and color photos of about 500 species of fossils and reconstructions of about 40 species; geological area maps. Intended for fossil collectors, amateurs, and students.

Western North America

1334. Bowen, Oliver E. **Rocks and Minerals of the San Francisco Bay Region**. Berkeley, CA: University of California Press, 1966. 71p. (California Natural History Guides, No. 5). $7.00pa. ISBN 0520032446; 0520001583pa.

18 "families" of rocks and minerals described; line drawings, some color and black-and-white photos; general geology; also features chart with distinctive characteristics of 105 rocks and minerals found in the area.

☐ Hall, Clarence A. **Natural History of the White—Inyo Range, Eastern California**. See entry 150.

1335. Leaming, S. F. **Guide to Rocks and Minerals of the Northwest**. North Vancouver, B.C.: Hancock House, 1980. 33p. $4.95pa. ISBN 088839053Xpa.

60 types; color photos; pamphlet.

1336. Ludvigsen, Rolf, and Graham Beard. **West Coast Fossils: A Guide to the Ancient Life of Vancouver Island**. Vancouver: Whitecap Books, 1994. 193p. ISBN 1551101491pa.

140 black-and-white photos; maps; sites for fossil hunting; much background and geology.

1337. Schreier, Carl. **A Field Guide to Yellowstone's Geysers, Hot Springs, and Fumaroles**. Moose, WY: Homestead, 1987. 96p. $12.95pa. ISBN 0943972094pa.

113 "thermal features" described and illustrated with color photos; arranged by location in Yellowstone; includes combined index/checklist.

1338. Tidwell, William D. **Common Fossil Plants of Western North America**. 2nd ed. Washington, DC: Smithsonian Press, 1998. 299p. $24.95pa. ISBN 1560987839; 1560987588pa.

350 species; black-and-white illustrations and photos, some color photos. For fossil collectors.

❑ Whitney, Stephen. **A Field Guide to the Grand Canyon**. 2nd ed. See entry 175.

❑ Young, Robert G., and Joann W. Young. **Colorado West: Land of Geology and Wildflowers**. Rev. ed. See entry 707.

Man-Made Objects

This chapter contains guides that identify man-made objects. Most follow the standard field guide format fairly closely. They are included for a variety of reasons. Some demonstrate how loosely the term "field guide" is used, others show how well the field guide format works for identifying manufactured objects, while still others identify man-made objects of interest to amateur naturalists.

NORTH AMERICA

1339. Aboulafia, Richard. **Collins, Jane's Civil Aircraft**. Glasgow: HarperCollins, 1996. 255p. $8.00pa. ISBN 0004709438pa.
         100 types; color photos; miniature (3x4 inches). Just one of a number of small Collins Jane's guides; others include *Warships of World War II*, *Aircraft of World War II*, *Tanks of World War II*, and *Combat Aircraft*.

1340. Baker, T. Lindsay. **A Field Guide to American Windmills**. Norman: University of Oklahoma Press, 1985. 516p. ISBN 0806119012.
         100 makes or models of windmills; black-and-white photos and silhouettes. Includes extensive history, comprehensive manufacturers' list, a very lengthy bibliography, in addition to a section identifying and describing windmills. This hefty volume is a good companion to take on drives through the country, but is not for the hiker.

1341. Brownstone, Douglass L. **A Field Guide to America's History**. New York: Facts on File, 1984. 325p. ISBN 0871966220.

Covers man-made features of the landscape, such as canals, bridges, buildings, wells, fences, and bottles that are often seen on walks or drives; discusses how to identify and date styles. Black-and-white illustrations.

1342. Butler, Joseph T. **Field Guide to American Antique Furniture**. New York: H. Holt, 1986. 399p. $19.95pa. ISBN 0805001247pa.

1,700 line drawings; covers the period from seventeenth century to the early twentieth century; has information on periods followed by illustrations of furniture arranged by type and period; no information on prices is given. Also published in 1985 by Facts on File. Illustrated by Ray Skibinski.

1343. Eastman Kodak Company. **Kodak Pocket Guide to Nature Photography**. 2nd ed. Rochester, NY: Silver Pixel, 1995. 112p. ISBN 0879850086.

Color photos and line drawings; includes information on topics such as blinds, lenses, subjects, habitats, etc.; pocket sized.

1344. Foster, Gerald L. **A Field Guide to Trains of North America**. Boston: Houghton Mifflin, 1996. 146p. $14.95pa. ISBN 0395701120pa.

170 types of locomotives and cars commonly found in North America; line drawings; grouped by visual similarity. The introduction specifically states that this guide is modeled after the Peterson's guides and it does, in fact, include "field marks."

1345. Gramly, Richard Michael. **Guide to the Palaeo-Indian Artifacts of North America**. Buffalo, NY: Persimmon, 1990. 76p. (Persimmon Press Monographs in Archaeology; Occasional Paper, Buffalo Society of Natural Sciences, No. 4). ISBN 0961546239.

150 types; black-and-white illustrations. Attempts to be a comprehensive listing of all types of paleo-Indian artifacts found in North America; arranged alphabetically; includes description and synonyms.

1346. Kaderabek, Todd. **A Field Guide to Hot Sauces: A Chilehead's Tour of More Than 100 Blazing Brews**. Asheville, NC: Lark, 1996. 132p. ISBN 1887374108.

100 types of hot sauces, including descriptions of tastes. Illustrated by Christi Teasley.

1347. McAlester, Virginia, and Lee McAlester. **A Field Guide to American Houses**. New York: Knopf, 1995. 525p. $23.95pa. ISBN 0394739698pa.

    39 styles from the colonial period to the present; pictorial key; black-and-white photos; arranged by period; background information. Illustrated by Lauren Jarrett and Juan Rodriguez-Arnaiz.

1348. Moldvay, Albert. **National Geographic Photographer's Field Guide**. Washington, DC: National Geographic Society, 1981. 120p. ISBN 087044395X.

    Spiralbound photography guide with tips on nature, landscape, night shooting, and existing-light photography; designed to be taken into the field.

1349. Montgomery, M. R., and Gerald L. Foster. **A Field Guide to Airplanes of North America**. 2nd ed., rev. and updated. Boston: Houghton Mifflin, 1992. 230p. $14.95pa; $22.95. ISBN 0395628881pa; 039562889X.

    350 types; black-and-white illustrations of side and overhead views; grouped by visual similarity. The introduction specifically states that this guide is modeled after the Peterson's guides and it does in fact include "field marks." Illustrated by Gerald L. Foster.

1350. Noble, Allen George, and Richard K. Cleek. **The Old Barn Book: A Field Guide to North American Barns and Other Farm Structures**. New Brunswick, NJ: Rutgers University Press, 1995. 222p. $29.95pa. ISBN 0813521734pa; 0813521726.

    Black-and-white photos and line drawings. Identifies 50 major types of barns and 42 types of other farm buildings, such as fences, cribs, or springhouses. Includes lists of where to find barn types, as well as what barn types to look for in an area. Based on volume 2, *Barns and Farm Structures*, of Noble's *Wood, Brick, and Stone*.

1351. Sherwood, Richard M. **A Field Guide to Sailboats of North America**. 2nd ed. Boston: Houghton Mifflin, 1994. 404p. $14.95pa. ISBN 0395652405; 0395652391pa.

    225 types of sailboats commonly found in North America; line drawings; grouped by visual similarity. The introduction specifically states that this guide is modeled after the Peterson's guides and it does in fact include "field marks."

1352. Simpson, Stephen J., and George C. McGavin. **The Angler's Fly Identifier**. Philadelphia: Running Press, 1996. 192p. $20.00. ISBN 1561386103.

100 artificial trout flies; color photos; angling tips. Includes section on trout behavior. See entry 828 for a guide to North American trout insects.

1353. Tully, Lawrence N., and Steve N. Tully. **Field Guide to Flint Arrowheads and Knives of the North American Indian: Identification and Values**. Paducah, KY: Collector, 1998. 175p. $9.95pa. ISBN 1574320165pa.

180 types of projectile points and 26 types of knives; black-and-white photos. Descriptions include state where found, size, and value.

1354. Whiteford, Andrew Hunter. **North American Indian Arts**. New York: Golden, 1983. Rev. ed. 160p. $5.50pa. ISBN 0307240320pa.

Color illustrations; covers pottery, textiles, etc. Illustrated by Owen Vernon Shaffer.

Eastern North America

1355. Cantor, Laurel Masten. **The Gargoyles of Princeton University: A Grotesque Tour of the Campus**. Princeton, NJ: Princeton University, Office of Communications/Publications, 1983. 36p.

A pamphlet illustrating 14 of the gargoyles found on Princeton University buildings and describing many more.

1356. Pillsbury, Richard, and Andrew Kardos. **A Field Guide to the Folk Architecture of the Northeastern United States**. Hanover, NH: Dartmouth College, Dept. of Geography, 1970. 99p. (Geography Publications at Dartmouth, No. 8).

A guide to the "housing of the common man" (from the foreword) in New England, Pennsylvania, and the Chesapeake Bay area. Black-and-white photos and line drawings; discusses the major house and barn styles of the eighteenth and nineteenth centuries.

1357. Turner, Ellen Sue, and Thomas R. Hester. **A Field Guide to Stone Artifacts of Texas Indians**. 2nd ed. Houston, TX: Gulf, 1993. 394p. (Gulf's Fieldguide Series). $29.95; $17.95pa. ISBN 0877192308; 0877192227pa.

200 types of dart and arrow points; black-and-white illustrations; maps; introductory materials. Illustrated by Kathy Roemer.

1358. Visser, Thomas Durant. **Field Guide to New England Barns and Farm Buildings**. Hanover, NH: University Press of New England, 1997. 232p. $19.95 (flexi). ISBN 0874517710 (flexi).

16 types of barns and 40 types of other farm buildings from New England; black-and-white photos and illustrations; arranged by form and use. The index includes locations of barns featured in the text.

Western North America

1359. Patterson, Alex. **A Field Guide to Rock Art Symbols of the Greater Southwest**. Boulder: Johnson, 1992. 256p. $15.95pa. ISBN 1555660916pa.

150 types of symbols; visual key; line drawings; covers the area from northern Mexico to Utah, and California to Colorado. Each type of symbol is described and its meaning discussed in a page or two (over 150 authorities were consulted). Also includes alphabetical list of ascribed meanings and descriptions.

Field Guide Publishers

Not all field guide publishers are included in this list, which would be an impossible task. An attempt has been made to include most of the North American small presses that publish several guides or that publish particularly notable examples. The major publishers that are found in most bookstores are not included.

All of the contact information was correct as of August 1998, but, of course, it is subject to change.

Alaska Northwest Books
P.O. Box 10306
Portland, OR 97210
(800) 452-3032

Publishes a number of books about Alaska's wild plants and animals.

Alaskakrafts
7446 E. 20th Ave.
Anchorage, AK 99504-3429
(907) 333-8212
fax: (907) 333-4989
e-mail: akkrafts@alaska.net
WWW: http://www.alaska.net/~akkrafts/

Alaskakrafts publishes several field guides to the plants of Alaska along with other local-interest books.

Blue Kirio
74-5602 Alapa St. No. 764
Kailua-Kona HI 96740
(800) 863-2524 or (808) 322-4317
fax: (808) 322-4021
e-mail: caseym@interpac.net
WWW: www.bluekirio.com/

This diving company publishes some interesting fish identification guides for the Pacific Islands.

Falcon Press
P.O. Box 1718
Helena, MT 59624
(800) 582-2665
fax: (800) 508-8638
e-mail:Falconbk@ix.netcom.com
WWW: http://www.falconguide.com/

Falcon publishes several field guides among its many outdoors/vacation titles. It is primarily a western publisher and also distributes books from other publishers. These include Flower Press, Waterford Press, and others.

Field Guides
32145 SW 202 Ave.
Homestead, FL 33030
e-mail: guides@icanect.net

Publishes a number of laminated card guides.

Flower Press
192 Larch Lane
Columbia Falls, MT 59912

Publishes several western flower guides.

Gulf Publishers
3301 Allen Parkway
Houston, TX 77019
(713) 529-4301
WWW: http://www.gulfpub.com

Gulf publishes several field guides, most covering Texan, Southwestern, or western topics.

Homestead Publishing
P.O. Box 193
Moose, WY 83012

Ibis Publishing Company
3420 Freda's Hill Road
Vista, CA 92084-7466
e-mail: ibispub@msn.com

Johnson Books
1880 South 57th Court
Boulder, CO 80301
(800) 258-5830

Publishes a number of books on adventure travel and natural history.

Lone Pine Publishing
16149 Redmond Way, #180
Redmond, WA 98052
(800) 661-9017
fax: (800) 424-7173)
e-mail: artdept@lonepinepublishing.com
WWW: http://www.lonepinepublishing.com/

Publishes a number of excellent field guides, most covering the western U.S. and Canada. Many of their guides are published both in the U.S. and Canada, with slightly different titles.

Massachusetts Audubon Society
(800) AUDUBON
http://www.massaudubon.org

Publishes several folded, laminated field guides dealing with local wildlife. The Massachusetts Audubon Society is not allied with the National Audubon Society.

Mountain Press
P.O. Box 2399
Missoula, MT 59806
(406) 728-1900 or (800) 234-5308
e-mail: mtnpress@montana.com

Publishes a nice array of popular western natural histories.

Mountaineers Books
1001 SW Klickitat Way, Suite 201
Seattle, WA 98134
(800) 553-4453
fax: (800) 568-7604 or (206) 223-6306
e-mail: mbooks@mountaineers.org
WWW: http://www.mountaineers.org/mbooks/mbcat.htm

Includes field guides among the mountaineering guides.

Natural World Press
47227 Goodpasture Road
Vida, OR 97488
(541) 896-0263
fax: (541) 896-0310
e-mail: russ@natworld.com

Publishes a variety of laminated cards and books, most featuring tropical reef fishes. Formerly of San Mateo, California.

Nature Study Guild
P.O. Box 10489
Rochester, NY 14610-0489
phone/fax: (716) 482-6090
(800) 954-2984

Publishes the Nature Finder series, among others.

Naturegraph
3543 Indian Creek Road
Happy Camp, CA 96039
WWW: http://members.aol.com/Naturgraph/homepage.htm

Publishes several Western and North American field guides and natural histories.

New World Press
1861 Cornell Road
Jacksonville, FL 32207
(800) 737-6558
fax: (904) 731-1188
WWW: http://fishid.com/

Publishes a number of snorkeler's guides.

Rainforest Publications
P.O. Box 193
Stanwood, WA 98292
phone/fax: (360) 652-1779
e-mail: rfp@whidbey.net
WWW: http://www.econiche.com/index.htm

Publishes the Costa Rica field guides series of laminated cards.

Ralph Curtis Publishing, Inc.
P.O. Box 349
Sanibel Island, FL 33957-0349
(941) 454-0010
fax: (941) 395-2727

Publishes or distributes a number of field guides, in particular the Photographic Wildlife Pocket Field Guide Series, which is published in the U.K. by New Holland.

Sea Challengers
35 Versailles Court
Danville, CA 94506-4454
(510) 327-7750
fax: (510) 736-8982

Publishes a number of books on marine invertebrates and fishes worldwide. Also distributes books on marine animals from other publishers.

Seahawk Press
6840 SW 92nd St.
Miami, FL 33156
fax: (305) 667-3572

Publishes a number of marine field guides, in particular plastic submersible guides to fishes and marine invertebrates.

Stackpole Books
5067 Ritter Road
Mechanicsburg, PA 17055-6921
(800) 732-3669
fax: (717) 796-0412
WWW: http://www.stackpolebooks.com/

Ten Speed Press
P.O. Box 7123
Berkeley, CA 94707
(800) 841-BOOK
fax: (510) 559-1629

Ten Speed Press includes some unusual field guides among its wide array of publishing ventures, including mushroom and humorous field guides.

Thunder Bay Press
2325 Jarco Drive
Holt, MI 48842
(517) 694-4616

Publishes a number of midwestern guides.

Timber Press, Inc.
133 SW Second Ave., Suite 450
Portland, OR 97204
(800) 327-5680
e-mail: orders@timber-press.com
WWW: http://www.timber-press.com

While primarily a horticultural publisher, Timber Press also offers several regional plant field guides and more technical floras.

Waterford Press
P.O. Box 4739
Blaine, WA 98231-4739
(360) 332-7301
fax: (360) 332-6084
WWW: http://www.waterfordpress.com

Publishes James Kavanagh's *Nature of* . . . and Pocket Naturalist guides (also distributed by Falcon Press).

Annotated Bibliography

This annotated bibliography is intended as a reasonably complete list of articles, books, and book chapters dealing with field guides. Only works that deal with field guides as an actual object or idea are included, since the term "field guide" has often been used more for its cachet or familiarity than for its reference to a particular type of identification guide.

Many articles have been written by individuals who know their field guides very well. These articles generally appear in natural history or birding magazines, and discuss field guides as a phenomenon. These natural history magazines are also the primary source for histories of the field guide genre (especially bird guides), and for biographies of the best-known field guide authors and illustrators. While several of Roger Tory Peterson's obituaries are included, no attempt was made to collect all of them. Only obituaries that contain significant amounts of information about the field guide genre are included.

Articles that review and compare several field guides or series are included in this bibliography, along with selected book reviews that discuss field guides in a more scholarly or philosophical manner. The numerous reviews of individual field guides are not included, but can be found in many sources from the *New York Times Book Review* to the *Quarterly Review of Biology*. Check almost any general-interest or book review index.

There do not appear to be any lengthy scholarly treatments of field guides, but several articles or books mention field guides as part of a literary or artistic spectrum, and are annotated and quoted in the bibliography. Poets, essayists, and other authors have taken inspiration from their relationships with their field guides, and several of the resulting works are also included. A number of poems are titled "Field Guide" but have nothing to do with field guides, and are thus not included.

Adler, Jerry. "The Original Field Guide." *Newsweek*, August 12, 1996, 60.
   A short biography of Roger Tory Peterson.

Appelbaum, Judith. "Pursuing the Wild Whatever." *New York Times Book Review*, September 25, 1983, 39–40.
   Provides sales figures and descriptions for Audubon, Golden, Peterson, and Simon and Schuster field guides.

"Armchair Birding: A Reading List." *Sunset*, February 1985, 70–72.
> Reviews the Golden, Peterson, and Audubon guides.

Elizabeth Bishop to Marianne Moore, 28 December 1941. *Elizabeth Bishop: One Art*, ed. Robert Giroux, 104. New York: Farrar, Straus, Giroux, 1994.
> "For Christmas Red gave me *A Field Guide to Birds* by R. T. Peterson. Have you ever seen it? I think it is marvelous and very different to most bird books: the pictures give the birds as they look at a distance and in all stages of age and molting, etc. . . . —the descriptive writing is quite good and different, too." Even poets liked Roger Tory Peterson's work.

Bond, Mary Wickham. *How 007 Got His Name*. London: Collins, 1966.
> The real James Bond's wife tells of the vicissitudes of life after Ian Fleming named his hero after her husband, the author of *Birds of the West Indies*.

Brooks, Paul. "A Field Guide to Roger Tory Peterson." *Blair and Ketchum's Country Journal*, December 1985, 34–41.
> A biography of Roger Tory Peterson; describes pre-Peterson identification sources and the genesis of the early volumes in the Peterson series.

Brown, Lauren. "Roger Tory Peterson: A Birder Looks to the Ground." *American Horticulturist*, August 1995, 32–37.
> A biography of Roger Tory Peterson, with emphasis on his work on the Peterson *Field Guide to Wildflowers*.

Cassie, Brian E. "Collecting the First Peterson Bird Guides." *AB Bookman's Weekly* 78 (1986): 355–356.
> Provides detailed descriptions and prices for the early editions of Peterson's field guides.

Caywood, Carolyn. "The Birds and the Bees." *School Library Journal* 43, no. 10 (1997): 61.
> Suggests that field guides can engage the interest of teens with short attention spans. Also discusses features that make up a good field guide, such as organization and illustrations (she does not favor photographs).

ng, Katherine S. "Field Guides for Insects." *Reference Services Review* (1984): 41–46.
> Reviews six major North American insect field guides.

"Choosing Your First Field Guide." *WildBird*, March 1990, 32–35.
> Five expert birders evaluate the top four field guides (Audubon, Golden, National Geographic, and Peterson)—and don't always agree.

Coletta, W. John, and Erik S. Munson. "Taxonomy, Ideology and the Structure of the Natural History Field Guide." *American Biology Teacher* 55 (1993): 456–462.

As you might guess from the title, this is not your usual field guide review. Criticizes most modern field guides for using a taxonomic rather than an ecological perspective in presenting information, thus they miss an opportunity to reinforce ecological thinking in students.

Connor, Jack. "David Sibley's Magnificent Obsession." *Audubon*, May/June 1998, 70–75.

The story of David Sibley, a master birder and artist who is working on a new field guide, *National Audubon Society Master Guide to Birds*. This two-volume set will consist of a field guide covering 800 species and a reference volume due in 1999. Sibley has already illustrated other bird books, including Dunne's *Hawks in Flight*. Some are calling Sibley the successor to Roger Tory Peterson's mantle.

Copple, Nathan. "A Field Guide to the Words . . . of Roger Tory Peterson." *Living* Bird, Winter 1996, 30–33.

While Peterson's innovations in the area of field guide illustration are well known, Copple attends to the more neglected aspect of Peterson's concise but often poetic descriptions and finds them delightful.

Culhane, John. "The Hobby That Lifts Your Heart." *Reader's Digest*, February 1987, 56–60.

Discusses how to get started in birding, including choosing binoculars, field guides, and bird song recordings. Includes an interview with Roger Tory Peterson.

Devlin, John C., and Grace Naismith. *The World of Roger Tory Peterson: An Authorized Biography*. New York: Times Books, 1977.

Besides being the biography of the best-known field guide author/artist, this book includes color prints of his art and discusses the history of several field guides in the Peterson series, not just the birds.

Dunne, Pete, Roger B. Swain, and Richard LeBlond. "A Guide to the Field Guides." *Natural History*, May 1987, 62–69.

Reviews the authors' favorite North American bird, mammal, tree, and wildflower field guides. Also lists a few regional field guides. All authors are experts, and even list the weight and size of the field guides.

Fengler, Jeff M. "An Expert for Your Pocket: Selecting a Field Guide to Suit Your Butterflying Needs." *Virginia Butterfly Bulletin* (Winter 1996): 6–7. Also available on the Web at http://www.baylink.org/vbs/butarts.html.

Describes and recommends butterfly field guides for a variety of purposes. Reprinted from *Connecticut Butterfly Association Newsletter* (October 1995).

"Field Guides for Better Birding." *Sunset*, October 1990, 47.
Describes the Peterson and Audubon Western bird guides.

Findlay, W. P. K. *Wayside and Woodland Fungi*. London: Frederick Warne and Co., 1967, 23–26.
In his chapter on the role of the amateur in mycology, Findlay tells of finding Beatrix Potter's paintings of fungi, which were never published in a scientific work during her lifetime. He went on to use 59 of these plates to illustrate this field guide.

Fisher, Alexander D. "Field Guide." *Grand Street* 14 (1996): 212.
There are several poems entitled "Field Guide," most of them having nothing to do with field guides or the natural world. While belonging to that category, this otherwise opaque poem at least shows an understanding of what a field guide is, since the author asks, "and have you read carefully the Field Guide to Hunted Beasts."

Foderaro, Lisa W. "Reluctant Earthling." *New York Times*, August 26, 1986, sec. C.
A biography of Roger Tory Peterson, with emphasis on his personal life in his later years.

Fothergill, Alastair. "Natural Classic." *BBC Wildlife*, October 1995, 81.
The author fondly recalls his favorite nature books, including Peterson et al. *Birds of Britain and Europe*; Fry et al. *Kingfishers, Bee-eaters and Rollers*; and Harris et al. *Macmillan Field Guide to Bird Identification*.

"Getting Serious About Wildflowers?" *Sunset*, March 1989, 82–83.
Reviews the Peterson and Audubon wildflower guides, plus several western guides.

Gollop, Bernie. "Review of a Butterfly Field Guide—80 Years Late." *Blue Jay*, September 1996, 152–155.
Reviews John Henry Comstock's *How to Know the Butterflies*, an identification guide first published in 1904. Gollop admires and exerpts Comstock's leisurely, descriptive comments. As Gollop concludes, "They don't write field guides like that anymore!"

Gordon, John Steele. "Inventing the Bird Business." *American Heritage*, December 1996, 18.
A short biography of Roger Tory Peterson.

Graham, Frank, Jr. "Mardy at 88." *Audubon*, November 1990, 12, 14.
> An interview with Margaret Murie, the widow of Olaus Murie, author of the Peterson *Field Guide to Animal Tracks* and herself a committed environmentalist. Also discusses the success of Olaus's field guide.

Graham, Frank, Jr. "James Bond." *Audubon*, May 1989, 12, 14.
> An obituary of the original James Bond, or rather, the author of *Birds of the West Indies* whose name was borrowed by Ian Fleming for Agent 007 because Fleming thought it sounded dull. Bond's work on the birds of the Caribbean can hardly be called dull, though.

Hadas, Rachel. "The Address Book and the Bird Book." *New England Review* 18 (1997): 123–124.
> The author muses over the meaning that her deceased mother's address book and annotated, dog-eared copy of Peterson's *Field Guide to the Birds* hold for her. "Both Peterson and my mother, as writers, give a sense of immediacy, of the truth of an experience. Peterson's meticulous transcriptions of bird calls, for example, always seem true to what he has heard, not what others have written. This is a guess of mine; but with my mother's book, I don't have to guess—if she saw a bird, she recorded it."

Hass, Robert. *Field Guide*. New Haven, CT: Yale University Press, 1973.
> The ex-Poet Laureate's first book of poetry. In this volume, many of the poems were inspired by nature, and two even refer to field guides as field guides, not metaphors for something else. In "Fall", the narrator and his friends go out mushrooming, and keep themselves awake at night remembering the symptoms of mushroom poisoning as described in their "terrifying field guide." In "Letter", the narrator tells his wife that her absence has taken away his enjoyment of plants and animals. He wanted to help his children identify a new flower, but quite apart from her absence taking the savor from the world, "You even have/my field guide."

Johnson, Cathy. "Nature's Pharmacy." *Country Living*, May 1997, 92–96.
> The author discusses the use of medicinal plants by early Americans and recommends several field guides to use in identifying plants and their uses. Very wisely, Johnson also includes warnings about using plants without being very sure you know what they are.

Kinch, Michael. "Field Guide to Trees." *Reference Services Review* 12 (Fall 1984): 35–40.
> Reviews 15 North American tree guides.

Klaas, Janet. "Birds of a Feather: A Covey of Field Guides." *Reference Services Review* 12 (Summer 1984): 27–40.

Reviews the National Geographic and Golden field guides to the birds, the eastern and western Peterson and Audubon guides, and the *Audubon Society Master Guide to Birding*. Includes sample text for the Purple Finch from all of the field guides.

ht, Frank. "Field Guides." *Conservationist*, December 1995, 31.

Brief reviews of the author's favorite bird, flower, tree, behavior, and habitat guides, plus a paragraph on Golden Guides.

Knopp, Lisa. "Field Guide." In Lisa Knopp, *Field of Vision*. Iowa City: University of Iowa Press, 1996, 120–124.

The author is first grumpy about her inability to identify a bird using a field guide, then philosophical about the power of names. "I have used field guides only to flesh out details about birds that other people have identified for me—my first purple finch, my first whippoorwill. Then I was struck by the differences between the guidebook model and the real thing, since plumage in the field is seldom as bright or as delineated as the sketched form; subjects are never as still, the light rarely as good, nor do the bird calls or songs match the descriptions. . . . The authors of field guides claim their books are not just for veteran birdwatchers, yet these experts seem to have forgotten how one bird can look and sound so much like the next to an untrained eye and ear (p. 121)." This sweeping condemnation seems primarily directed at the Golden Guide *Birds of North America*, but once she identifies her mystery bird (a killdeer), it suddenly seems much more like its depiction.

Lane, Margaret. *The Tale of Beatrix Potter*. Rev. ed. London: Frederick Warne, 1968.

Beatrix Potter may not be the first name to leap to mind when thinking of field guide artists, but before she published *The Tale of Peter Rabbit* she created many watercolors of fungi that were posthumously published in Findlay's *Wayside and Woodland Fungi*. Her biography details Potter's disappointment at not being taken seriously in her youth.

Larson, Gary. *There's a Hair in My Dirt!* New York: HarperCollins, 1998.

Papa Worm tells the story of Harriet's walk in the woods as she blithely misunderstands the natural world. Among the events she overlooks are bears examining their *Field Guide to the Humans* (including varieties such as Logger, Hunter, Mushroomer, and Elvis).

Leo, John. "Birds: A Guide to the Guides." *Discover*, June 1984, 96.

Evaluates the six most common bird field guides for North America (Audubon, Golden, National Geographic, and Peterson;

also the *Audubon Society Master Guide to Birding* and Pough's three Audubon guides).

Leo, John. "He was a Natural." *U.S. News and World Report*, August 12, 1996, 17.
    A short biography of Roger Tory Peterson.

Line, Les. "Wildlife by the Book." *National Wildlife*, April/May 1995, 46–53.
    Talks about the illustrators of some popular Peterson field guides: Amy Bartlett Wright (*First Guide to Caterpillars*), Vera McKnight (*Mushrooms*), Richard White (*Beetles of North America*), John Douglass (*Atlantic Coast Fishes*), Janet Wehr (*Eastern Trees*), and, of course, Roger Tory Peterson (*Birds*). Lavishly illustrated with examples of the illustrators' field guides.

Lipske, Michael. "He Wrote the Book on Birds." *National Wildlife*, December/January 1996, 66.
    A short biography of Chandler S. Robbins, author of the Golden Field Guide *Birds of North America*.

Lipske, Mike. "They're Not Called Birdwatchers Anymore." *National Wildlife*, October/November 1986, 46–51.
    A light-hearted look at birders and birding, including interviews with Roger Tory Peterson and a discussion of the popularity of field guides.

Lyon, Thomas J. "A Taxonomy of Nature Writing." In Lyon, Thomas J., ed. *This Incomperable Lande: A Book of American Nature Writing*. Boston: Houghton Mifflin, 1989, 3–7.
    Field guides are placed in a spectrum of nature writing ranging from presentation of natural history information (field guides and scientific papers) to personal responses to nature (Annie Dillard's *Pilgrim at Tinker Creek*) to philosophical interpretations of nature (any of John Burroughs' work). "If conveying information is almost the whole intention, for example . . ., the writing in question is likely to be a professional paper or a field guide or handbook, most of which are only intermittently personal or philosophical and also, perhaps, literary only in spots (p. 3)." On the other hand, Lyon mentions at least one place in Peterson's *A Field Guide to Western Birds* that is particularly evocative.

Lythgoe, Michael H. "Field Guide." *South Dakota Review* 34 (Winter 1996): 74–75.
    Early fall wildflowers of Virginia, poetically rendered.

McWilliams, Patricia. "A Field Guide to Field Guides to Birds." *Country Journal*, February 1989, 14–15.
    Brief reviews of the Audubon, Golden, and Peterson bird guides.

Page, Jake. "A Golden Guide of Memories." *National Geographic Traveller*, November/December 1989, 16–18.

> The author waxes eloquent about the memories invoked while looking at the notes he made in his beat-up old Golden guide to the birds, though when the poor thing fell apart he switched to the National Geographic guide. The article is not so much about the virtues of field guides as the joys of birdwatching while travelling.

Peterson, Roger Tory. "Books of a Feather." *National Wildlife*, December 1983/January 1984, 22–28.

> The man himself traces the history of field guides. Lavishly illustrated with classical images from major bird illustrators.

Potts, Lesley S. "Guides for the Wildflower Pilgrim." *Reference Services Review* 13 (spring 1985): 29–35.

> Reviews 20 regional and national wildflower field guides for North America.

Rosen, Jonathan. "Why I Am for the Birds." *New York Times Book Review*, June 1, 1997, 55.

> Rosen discusses the joys of birdwatching, field guides and their literary merits, and the relationship between nature and literature (all in one page!).

Schalaway, Scott. "Complete Guide to Field Guides." *Pittsburgh Post-Gazette*, March 22, 1998, sec D.

> Despite the title, this is not a *complete* guide to field guides. The author recommends field guides for beginners in the eastern U.S., selecting from among the most popular series. He does cover most subjects, from wildflowers to night sounds.

Schmidt, Diane. "Field Guides for Everyone: A Guide to Their Diversity." *Reference Services Review* 21 (fall 1993): 43–48, 61.

> A general overview of the types of field guides that are available.

Seton, Ernest Thompson. *Two Little Savages; Being the Adventures of Two Boys Who Lived as Indians and What They Learned*. New York: Doubleday, Page and Co., 1903.

> In the chapter "How Yan Knew the Ducks Afar," the hero, having only guides that identified birds in the hand, sketched stuffed ducks as they would look from a distance. Seton, a well-known illustrator himself, included sketches and brief descriptions of what his character described as the "little blots and streaks that are their labels." This idea of simplified illustrations and descriptions was one of the major influences on Roger Tory Peterson's first field guide, and in fact, the two sets of duck drawings are very similar.

Tennent, W. J. "Higgins & Riley—the European View." *Entomologist's Gazette* 43 (1992): 257–261.

> Compares the British, French, and German editions of the *Field Guide to the Butterflies of Britain and Europe*, and finds a number of differences in the taxonomy and arrangement of the three versions.

Thomas, Francis-Noel, and Mark Turner. *Clear and Simple as the Truth: Writing Classic Prose*. Princeton, NJ: Princeton University Press, 1994, 115–117.

> The authors laud the prose in *The Audubon Society Field Guide to North American Birds, Eastern Region* as an exemplar of clear, classic writing that does not draw attention to itself. "A field guide, in its stand on truth, presentation, scene, cast, thought, and language, fits the classic stand on the elements of style perfectly. Its implied model is one person presenting observations to another, who is in a position to verify them by direct observation (p. 116)."

Thomson, Keith Stewart. "Bird-watchers Watched." *American Scientist* 81 (1993): 222–225.

> A history of field guides, with emphasis on pre-Peterson bird identification tools; also includes a discussion of "jizz."

Tufte, Edward R. *Visual Explanations: Images and Quantities, Evidence and Narrative*. Cheshire, CT: Graphics Press, 1997, 114–116.

> In this book dealing with good and poor ways of presenting information, the author lauds several submersible fish field guides as models of good design. Examples are taken from the *Waterproof Guide to Corals and Fishes of Florida, the Bahamas, and the Caribbean* by Idaz and Jerry Greenberg and Paul Humann's *Reef Fish Identification: Florida, Caribbean, Bahamas*.

Willis, Monica. "Fall Forays." *Country Living*, September 1997, 104.

> Reviews the Golden and Audubon Society field guides to trees.

Wilson, Mark. "How to Choose Your Field Guide Wisely." *Boston Globe*, June 1, 1997.

> The author recommends several field guides for beginning birders, coming out in favor of the Peterson guide, followed by the National Geographic guide, and the Stokes field guide. He is generally anti-photos.

Wolkomir, Richard. "In Deepest Gotham, Stalking the Wild and Woolly Gargoyle." *Smithsonian*, August 1993, 86–92.

> Mentions plans of the American Association for the Advancement and Appreciation of Animals in Art and Architecture (AAAAAAA) to publish a field guide to the architectural animals found on New York City buildings. The AAAAAAA hopes to finish this project around 2000.

"You Need a Field Guide—Maybe Two or Three." *WildBird*, August 1993, 50–53.

Evaluates the five "adequate" field guides for birders: the Audubon, Golden, National Geographic and Peterson guides, plus the *Audubon Society Master Guide to Birding* and the Peterson guides to bird nests.

Author/Illustrator Index

This index lists field guide authors and illustrators. Reference is to entry number. The letter "n" designates citations to authors in annotations. The letter "p" designates page numbers.

261

Title/Series Index

This index includes the titles of individual volumes and series, which are subdivided by chapter. Lists of audio, multimedia, video, and humorous guides are also provided. Reference is to entry number. Numbers in bold typeface refer to series descriptions. The letter "n" designates citations to titles in annotations, and the letter "p" refers to page numbers.